7.19.96
$42

LIGHTING

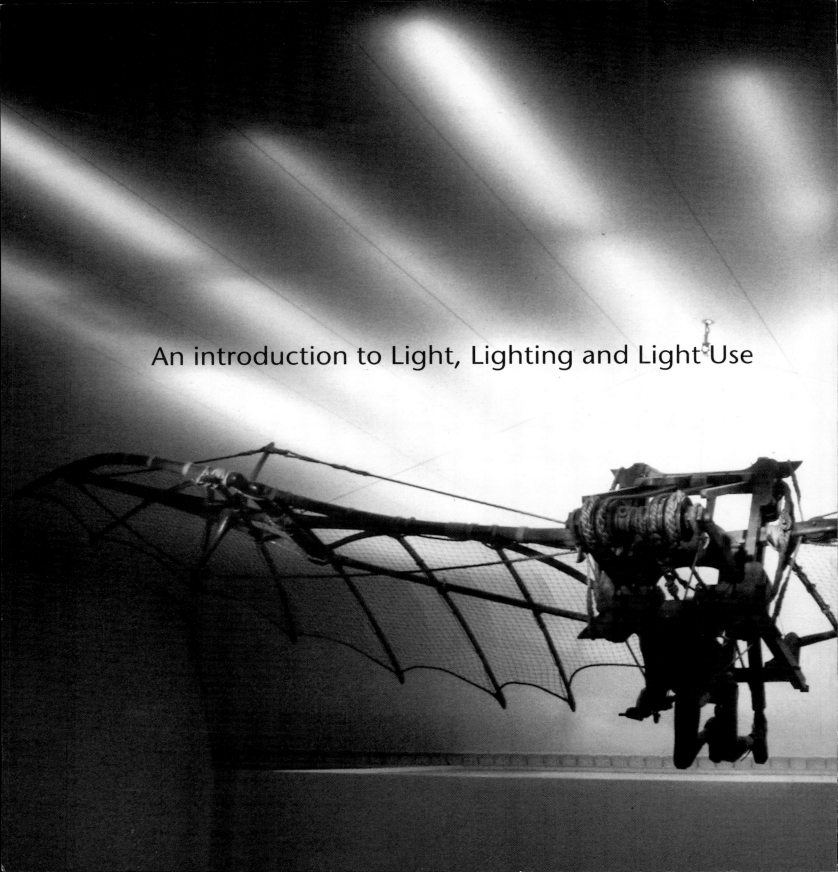

An introduction to Light, Lighting and Light Use

Janet Turner

LIGHTING

B.T. Batsford Ltd, London

Edited by
Conway Lloyd Morgan

Designed by the Armelle Press
Cover shows the Sainsbury Centre at the University of East Anglia (architect: Foster Associates, lighting consultant George Sexton)
Typeset by Wandsworth Typesetting
Printed in Hong Kong

A CIP record for this book is available from the British Library

for the publishers
B.T. Batsford Ltd.
4 Fitzhardinge Street
London W1H OAH

ISBN 0 7134 7292 8

CONTENTS

The Royal Society of Arts, in London, is a centre where the arts, industry and design meet. Housed in an eighteenth-century building, the RSA nonetheless has a contemporary attitude, and the lighting scheme is intended to emphasize this, both on the exterior (right) and in the restaurant and bar in the vaults (facing page).

Beginning in the middle

This is a book about lighting design, and so I make no apology for starting with design and the design process. Many books begin with the physics and physiology of light, then talk about lighting fixtures, luminance and other technicalities before getting down to cases. Because lighting is about so much more than technicalities, because with lighting the whole is larger than the sum of its parts, I want to begin by talking about the design process: the technical material and more detailed explanations will follow. And the explanation of any technical terms I use here will be found in the later sections of this book.

The Royal Society of Arts

One day in 1989 I was delighted to get a phone call from Christopher Lucas, secretary of the Royal Society of Arts in London. He was looking for advice on lighting their building in John Adam Street. There the architect Sam Lloyd was engaged on redeveloping part of the historic building. The RSA had been founded in the eighteenth century 'for the encouragement of Arts, Manufactures and Commerce' as its full title puts it. The Society advises on education and training, and runs a programme of meetings, conferences, exhibitions, awards and scholarships on aspects of its interests. A worldwide membership is drawn widely from industry, design and fine art, and its London home provides a focal point for its various activities. In particular, the RSA have for a long time insisted on more contacts between design and industry. I have always been involved on both sides in this debate, and

INTRODUCTION

The use of space at the RSA - for meetings and for leisure - needed different but coherent lighting solutions. In the vaulted part of the building this was generally uplighting (as in the restaurant, above, and cafeteria, facing left) but could be supplemented with downlighting, for example in the lecture theatre (facing)

long convinced that the closer the two could be the better. The invitation to work with the Society was therefore very welcome.

The RSA's building is a modest but imposing four-storey Georgian house, near the Thames on the western edge of the City of London. The architect's brief was to design a new junction between two parts of the building, creating a new staircase and entrance area, and a new lecture theatre. In particular he was also to redevelop the old wine cellars in the basement to provide a restaurant and bar for members, which could also be used as additional exhibition and conference space.

The first thing I did was to go with the architect to look around the building, the second, but equally important, was to look at his plans and discuss his ideas. It was clear at once to me that any lighting solution must not impose on the fabric of the historic building. Rather, the lighting should complement the architecture. For example, in the vaults the lighting could be used to emphasize the barrel-vaulted ceilings of the new bar and restaurant, to provide a relaxing atmosphere, while in the entrance hall somewhat more dramatic lighting could be used to make a formal statement of welcome, to frame the visitor's arrival. As the restaurant doubled as a conference area, flexibility was the keynote, to allow for different kinds of activities - formal lectures or round table discussions, slide shows or video presentations, exhibitions and so on.

From these discussions and visits we were able to define a brief for the job, and from the brief to make outline plans of the spaces to be lit, and look at possible solutions. Emergency lighting facilities had to be incorporated into the system, of course, and

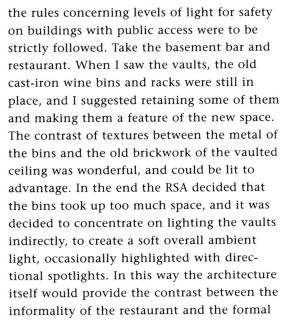

the rules concerning levels of light for safety
on buildings with public access were to be
strictly followed. Take the basement bar and
restaurant. When I saw the vaults, the old
cast-iron wine bins and racks were still in
place, and I suggested retaining some of them
and making them a feature of the new space.
The contrast of textures between the metal of
the bins and the old brickwork of the vaulted
ceiling was wonderful, and could be lit to
advantage. In the end the RSA decided that
the bins took up too much space, and it was
decided to concentrate on lighting the vaults
indirectly, to create a soft overall ambient
light, occasionally highlighted with directional spotlights. In this way the architecture
itself would provide the contrast between the
informality of the restaurant and the formal

INTRODUCTION

9

Georgian room spaces upstairs. The Adam architecture did not mean using period fittings, which would also have been out of step with the contemporary aims of the Society, nor would they have given sufficient flexibility.

The new staircase was the focal point of the architectural work, and so this needed to be assured by the way it was lit. So in one space - the basement - we opted for uplighting, letting the light be reflected off the ceiling brickwork indirectly, and in the other we used downlighting to spotlight the structural details of the staircase. The same logic influenced the detailed choice of lamps used, to achieve different colour of white light in the two spaces, softer and warmer below, brighter above. Next came the choice of fittings that would blend in with the architecture in both cases, and particularly in the basement involve as little interference as possible with the fabric of the walls.

The final design solution for the vaults was to insert uplighters into the floors in front of the pillars supporting the cross vaults, supplemented by an unobtrusive suspended track combining uplighters on the top and spotlights below. On the stairway low voltage spots were mounted under each half-landing to spotlight the baluster rail and treads. This lighting was supported by general purpose wall-washers on each side of the doors in the surrounding walls.

At all stages in the process ideas, plans and samples were discussed with the architect and the clients. In fact the RSA set up a design committee to oversee all aspects of fitting out the new design, including the lighting. Not surprisingly, the committee members, several of whom came from the design profession, contributed a lot of input

The staircase at the RSA is particularly important: it is lit by miniature spots under the half-landings, supplemented by wall-mounted lights (facing page). The use of uplighters on the old brickwork in the restaurant (above) is particularly effective.

Nicholas Grimshaw &
Partners' Pier 4A at Heathrow airport (above) was a
contemporary structure -
some would call it high-tech
- with a curved roof which
needed to be matched by a
similarly innovative light
(facing, below). The version
in the drawing (facing,
above) is the standard unit:
forty-two other versions were
produced for the scheme.

into the design decisions. The challenge of
meeting this range of different approaches
reminded me yet again that the right lighting
solution is often a matter of balance and
compromise, and only really works if all the
requirements of the space to be lit, both from
the clients' side and the designer's, are taken
into account.

Pier 4A, Heathrow

A year or so after completing the RSA scheme
Concord was asked to advise on the lighting
of the new Pier 4A at Heathrow airport. This
was a completely new building, and its
nature and intended use posed a range of
different problems. Pier 4A was designed to
provide additional passenger terminal space
at one of the world's busiest airports, particularly to serve domestic flights and flights to
the Channel Islands and Eire.

Nicholas Grimshaw and Partners were
the architects, and their uncompromisingly
contemporary design provided for passenger
movement on an upper level with vehicle
traffic and service access below. The passenger area, with its arched ceiling, curved walls
and serpentine plan, was based on the layout
of an aircraft interior, including design
features such as rounded corners on the
windows. Concord won the tender to provide
the lighting system, and it was at once clear
to us that such a novel building - largely a
single tunnel over a kilometre long - required
a novel solution. While conventional fittings
were used in the passenger lounges and duty-
free shops, for the long central corridor we
designed a special wing-shaped unit, suspended from the ceiling on struts.

This aerofoil section carries main and
emergency power supplies, cabling and

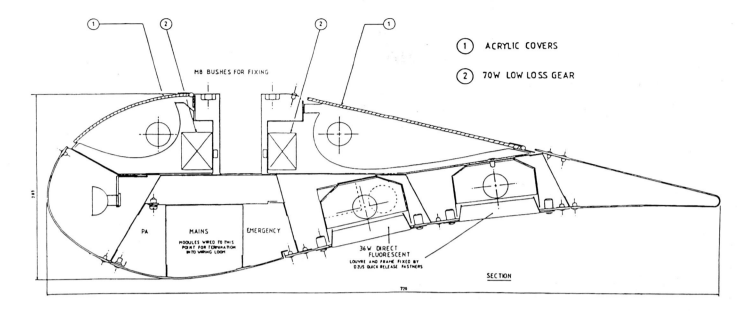

1 ACRYLIC COVERS

2 70W LOW LOSS GEAR

M8 BUSHES FOR FIXING

PA

MAINS

EMERGENCY

MODULES WIRED TO THIS
POINT FOR TERMINATION
INTO WIRING LOOM

36W DIRECT
FLUORESCENT

LOUVRE AND FRAME FIXED BY
DZUS QUICK RELEASE FASTNERS

SECTION

public address systems as well as lighting. Most wing units - normally 1.9 metres long - contain twin tube 70 watt fluorescent uplighters to wash light over the ceiling, with additional 3 watt fluorescent downlighters to provide a total average of 250 lux at floor level. Along the front edge of each wing 20 watt low voltage metal reflector lamps are fitted, behind curved glass fittings. These pick up the underlying design theme by emphasizing the wing shape and echoing an aircraft's landing lights. A large number of variations on the basic fitting was needed, for example to provide additional downlighting over passenger gates, or to modify the overall length of the unit to fit the building's complex design.

Creating a special lighting unit for a particular application is always an exciting challenge. In the case of Pier 4A we were also part of the design team on an important new

INTRODUCTION

13

The lighting wing shows the integration of form between it and the surrounding architecture (facing and above). More conventional lighting was used in the passenger lounges (right).

building, in a very visible site which in turn was subject to considerable constraints in terms of access, security, passenger flow and safety. The simple part of the solution - using an aerofoil shape as the basic design - emerged from talks and site visits with architect and client. The complex part was fitting in appropriate, safe and elegant lighting, as well as the other services to be carried. (For example the emergency lighting units and public address units were designed to fit into identical spaces, for flexibility.)

Details such as the finish of the wing and the positioning of the 'landing lights' needed to be decided during regular meetings and presentations, as did the development of the single strut attaching the units to the ceiling. The initial design ideas were presented in a full size mock-up of the final design, so that appearance and performance could be tested before the final specification was agreed.

These two projects give, I hope, some idea of the range of problems a lighting designer has to meet. Part of the pleasure of the job comes from the variety of spaces to be lit and the wealth of technology with which to achieve that, much from the challenge of meeting different requirements within a single solution, and much from the opportunity to work alongside architects and interior designers in meeting the needs and wishes of client and public. This book explores all these areas, and is intended to help designers, architects and clients better understand the potential modern lighting has.

The first chapter looks at the physics and physiology of light, which is one of the most important energy sources known to us. We then turn to the basics of lighting, in the second chapter, setting our a basic technical

armoury for the designer. In the third chapter we look at the applications of light, turning in the fourth chapter to the design process itself. A number of projects are analysed in more detail, including the use of computers for lighting design, and in the fifth chapter we look at a range of lighting solutions for different environments and uses. A final section contains the technical material and further reading.

Janet Turner
London 1994

WHAT IS LIGHT?

Light as energy

We all think we know what light is: it is the form of visible energy we get from sunlight, or from a candle flame, or from an electric lamp. Light, either directly from a light source or reflected off an object, is perceived by our eyes, and analysed into images in the brain. "Normal" light is white, but we know that it consists of different colours, the colours of the spectrum, which we see most commonly in the rainbow, or in the science laboratory when light is passed through a triangular glass prism. We probably remember that the speed of light, 186,000 miles per second, or 300,000 kilometres per second, is the fastest speed known to science, and that that value plays a part in Einstein's famous equation, $e=mc^2$. We think in terms of rays of light, and light waves. And we use and appreciate light every day: but what is light?

This seemingly simple question has baffled thinkers and scientists for centuries. In the early eleventh century AD the Arab mathematician Alhazen developed the notion of light as a form of energy working in rays (as the Greek thinker Euclid had suggested), but turned round Euclid's view, in which the rays of light came from the eye, to show that the eye collected light reflected by objects. Seven hundred years later, in 1704, Sir Isaac Newton published his book **Opticks.** In this he discussed his theory of coloured light, based on his experiments with a prism, and developed further his notion of light rays, defining light energy as being made up of streams of particles. At the same time the Dutch scientist Christian Huyghens was defining light as energy in wave form - a theory that offered a better explanation of the refraction effect of light Newton had analysed with his prism. The development of these two ideas of light - particle or wave - in the twentieth century has marched in parallel with the development of science in general, for in many ways thinking about light has stimulated scientific thinking in other areas, particularly nuclear physics, quantum theory and relativity.

We know now that light is a form of electromagnetic energy, like radio waves, but is light a form of wave energy or of particle energy? Sometimes it behaves like a wave form, sometimes as a particle. Famous experiments to decide between the two, by Thomas Young and others later, have only led to the conclusion that light can behave both as a particle and a waveform. This is termed the particle-wave duality of light, that is, the notion that light is both a particle form of energy, and a wave form: it exhibits the behaviour of both. The photon, as the unit of light energy is called, is an equally bizarre object: it has velocity but no mass - it literally exists only in motion.

The lighting designer does not, of course, need to know about the particle-wave duality of light for practical work, but I think it is valuable that the designer has some idea about the scientific importance of light and its place in modern thinking. The apparent illogicalities in the nature of light are a metaphor for the complexities of modern scientific theory; indeed the playwright Tom Stoppard used the particle-wave duality as a running metaphor in a recent play about the world of espionage.

The practical aspects of light in use do draw on the fact that light is a paramount form of energy. Its special nature means that light does not appreciably lose energy with distance, and its wavelike behaviour means that light reaches out in all directions evenly and in straight lines from a light source, until it

(Facing) A drawing by Goethe illustrating his theories of light.

Changing space with light

These pictures show how our perception of a space can be changed by the way it is lit. The interior of Westminster Cathedral, in London, is partially lit by daylight from the clerestory windows. Adding different lighting effects changes the impression and volume of the space. A simple downlighter (right) illuminates the altar, while a direct light into the area (far right) from above highlights it. Alternatively (facing, left) the altar can be backlit for greater relief.

The level of light in the nave can be varied (far right, top and bottom) to change the mood as well as the luminosity.

meets an obstruction. But light is only one form of electromagnetic energy. By light we mean those wavelengths of the total range of electromagnetic energy that are visible to the human eye. This represents a range from about 350 nanometres (one nanometre is one thousand millionth of a metre) for purple to over 700 nanometres for red. Red is the longest wavelength of visible light (in Newton's words it has the "least refrangible rays"): longer wavelengths of electromagnetic energy occur with infrared waves, and shorter ones than visible light with ultraviolet rays, gamma rays and X-rays.

While we often describe light in terms of colour, we should look at the terms used to describe white light before turning to a discus-

sion of colour. When we talk of a light being too bright or too dark, or of a room being overlit or underlit, how are such effects to be measured and accurately described? The first thing to make clear is that there are two different concepts here. One is the light source itself, and the other is the effect the light creates in a space. The first is independent of the second: the same lighting unit may be right for lighting a dining-room but be useless in an operating theatre. Turning to light sources first, the power of a light source is defined in terms of its luminous flux, more narrowly defined as its luminous intensity. The luminous flux is measured in lumens: they are measurements of the flow of light. The older term for candela, candlepower, is still used in

the United States of America: it gives an idea of the light value involved, even if it is old-fashioned. In modern terms a conventional 100 watt tungsten filament lamp gives out about 1,200 lumens, for example, a 500 watt tungsten halogen lamp gives out 9,500 lumens, and a 1500 millimetre 65 watt fluorescent tube about 5,000 lumens.

Looking at the measurement of light in a space, the term lumen, as a unit of light flow, can also be used to discuss the light received by a surface, reflective or not. This is called illuminance, meaning the luminous flux falling on a surface. An older term for this was the level of illumination or the illumina-tion value. This value is measured in lux, one lux being the illumination given by one

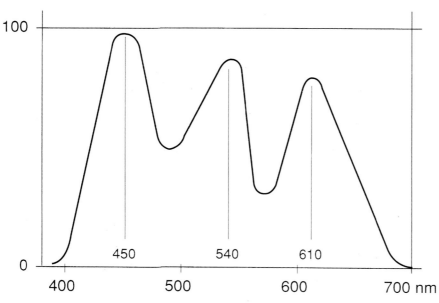

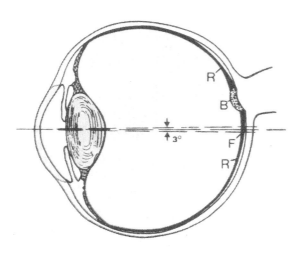

This diagram shows the different wavelengths of light visible to the human eye, measured in nanometres. Ultra-violet and infrared light are at the opposite, invisible, ends of the spectrum.

Diagram of the human eye

lumen over one square metre. The American expression for this, not based on the metric system, is the foot-candle (i.e. one candlepower over a square foot of area.) As a rule of thumb, therefore, one foot-candle is ten lux. The level of illuminance required varies according to the use the space illuminated is put to: the recommended illuminance for a theatre auditorium is 100 lux, for example, while for an airport check-in desk (where tickets and passports may be handled) it is 500 lux (as we saw in the case of Pier 4A earlier), and for colour-matching in a chemical laboratory 1,000 lux is suggested.

Where the light is emitted by a reflected source, it can be defined in terms of luminance, that is the luminous flux of the area, expressed in candela per square metre. The quantity of light reflected by a surface is also an important consideration: this is called the reflectance of a surface, and it is normally defined as a fraction of the illuminance: so that if a surface, for example a desktop, reflects back 300 of the 900 lumens falling on it, it can be said to have a reflectance of 0.3, or 30 percent. This reflectance is very often a quality of the material the surface is made of, or the paint finish on it. A glass mirror has a reflectance of about 97 percent, for example, while matt black cloth has a reflectance of 7 percent. A plain plaster finish reflects about 80 percent of incident light, a brick wall about 25 percent.

Tables of reflectance, and of recommended illuminance levels, are published by standards institutions. In the appendix to this book you will find some of the basic formulae for converting between the different units, and the terms used here are also listed in the glossary: the lighting designer needs to understand the basic terms and their relationships with each other.

Colour

Turning now to the question of colour, the main colours in the visible spectrum are red, orange, yellow, green and blue (in wavelength order) though in fact the spectrum presents a continuously changing range of colours - there is no fixed barrier between red, orange and yellow, for example, rather a continual progression of colour from hue to hue. The primary colours, in light terms, are green, red and blue. With evenly balanced light sources of these three colours, any of the other colours of the spectrum can be produced by adding light to light. Red and green light produce yellow, red and blue magenta (a crimson/purplish colour), blue and green cyan (a greenish blue). All three colours added together produce white light.

When we see red light, for example on a traffic light, what we are seeing is white light (from the light source) from which the other colours have been filtered, only allowing the red light to pass through. Technically speaking, the filter absorbs the light of the wavelengths other than those in the red part of the spectrum. The only true coloured light sources (where the light leaving the source is of a single wavelength, as opposed to being filtered or refracted by a prism) are found in lasers, which are designed to produce narrow coherent beams of monochromatic visible light or infrared radiation.

When we see a coloured object, however, what we are seeing is not coloured light but reflected light. The object absorbs the light of other wavelengths than red (or blue, or orange) and reflects back the coloured ones. We therefore see the object as coloured, not because it is emitting a coloured light, but because it is reflecting only certain wavelengths of white light. A white object,

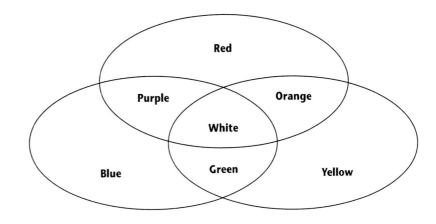

The primary colours can be combined additively with light to produce other colours in the spectrum, as here.

The effects of light direction

The direction of light falling on an object changes how we see it. In these examples, the same objects have been photographed, using light of a similar colour. On this page a bunch of twigs are lit from the front and from below: on the facing page the same subject is lit from the side. The different placing of the light changes the apparent shape and colour of the object, as well as affecting our perception of its depth.

On the facing page the broomstick is lit from the side by a spotlight (top) and a floodlight (below). Note how the different distribution of the light beam affects the appearance of the object. Below the broom is uplit, but in the left hand image the light is from the front, and in the left from behind the object.

WHAT IS LIGHT?

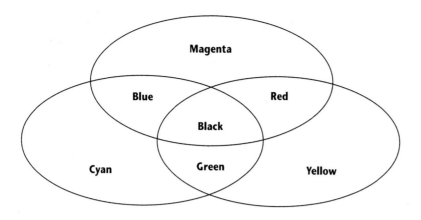

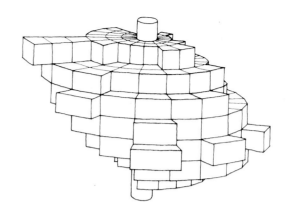

Subtractive colour effects are created when pigments, rather than light, are mixed: the total effect is of black, rather than white, and a different set of complementary colours are produced (right).

The Munsell colour guide (far right) defines colours according to their hue, value and chroma.

therefore, reflects all wavelengths, and a black object absorbs all wavelengths. Two things follow from this. Firstly, if the light illuminating an object is not white itself, the object will not necessarily appear the same colour as under white light, as all the wavelengths it seeks to reflect may not be present.

A good example of this is the low pressure sodium light used to illuminate motorways at night. This gives an orangey yellow light, in which many objects not in that part of the spectrum simply appear grey or black. The second point concerns mixing colours through pigments or paints rather than through light, in other words using reflected rather than direct colour. Here a subtractive process applies, not the additive process of mixing colours with light. The primary colours in pigments are therefore yellow, cyan and magenta. Mixing yellow and magenta produces red, mixing yellow and cyan produces green, and mixing cyan and magenta produces blue. Mixing all three together does not produce white, but black. One application of this principle is here on the page. If we look at a colour illustration under a magnifying glass, we will see that it

is made up of cyan, magenta and yellow dots (with black dots as well for added contrast) which through varying densities produce the visual effect of a range of different colours through a four-colour printing process. Red, green and blue light, on the other hand, are the sole colours used in creating television images.

The different appearances of different colours do not depend entirely on the mixture of pure colours, however. The lightness or darkness of a colour (its brightness or value) is also an important factor, as is its degree of saturation (or chroma). The hue may be seen as the colour value, the saturation (or chroma) the degree of that colour value, and the brightness (or value) the extent to which the colour value, whatever its degree of saturation, has been brightened or darkened.

There have been many attempts to fix exact values for different colours on the basis of these qualities of colour. Perhaps the best known and most widely-used system is that first proposed by A.H. Munsell in 1898. He grouped colours in a series of layers radiating out from a central spine. The overall effect of

Munsell's colour solid is of a distorted sphere, which can be cut into to find a specific colour and define its hue, value and chroma. The values for chroma and value radiate out from the spine (for chroma), and up and down the spine (for value). These qualities of colours can be identified numerically, by a system in which value is ranged from zero (black) up to ten (white) and chroma from zero (a neutral grey) to sixteen (the colour at full strength). A colour can be specifically identified by its Munsell number, which consists of an opening set of numbers and letters for the hue, followed by the figures for value and chroma. 10BG7/3, therefore, is one of the blue-green colours (BG) with a value of seven and a chroma of three.

    An explanation of what light and colour consists of is necessary for a designer's understanding, but being aware of what light means may be more relevant. Light does not merely have a technical dimension, important though this is for the designer. Light also has physiological and psychological aspects, of which the designer needs to be particularly aware.

Perception

From a physiological point of view, most of the sensory data human beings acquire about the world about them comes through their eyes. In Professor Richard Gregory's words "we are linked by light to objects." This is not to devalue the sensory data acquired through touch, taste, smell and hearing, but it does establish the primacy of vision as a means of acquiring data. Within the eye itself, a lens focuses the received light onto the retina at the back of the eye. Attached to the retina are thousands of tiny receptors, sensitive to light, that send a message to the brain when activated. Some of these sensors, called rods, react to white, grey and black. The others, called cones, are stimulated by red, blue and green light (thus confirming the primacy of those colours). Like the image on a television screen built up from small pixels (picture cells) of information, that are each a single colour, so the sensory centres in the brain construct a picture from the stream of individual bits of information from the rods and cones.

    In terms of evolution, the human eye is an extremely important feature. The fact that our eyes sit in the same vertical plane makes binocular vision, and so the ability to judge distances, possible (a bird's eyes, for example, do not sit in the same plane: a bird has a wider field of vision without the binocular overlap.) The diameter of a rod or cone is about that of the average wavelength of light: smaller or larger rods and cones would therefore be less effective, and a smaller eye would simply hold fewer of them, and so be less effective in evaluating information (a mouse's eye, for example, is smaller, and the wee cowering beastie's vision consequently less acute.) It has also

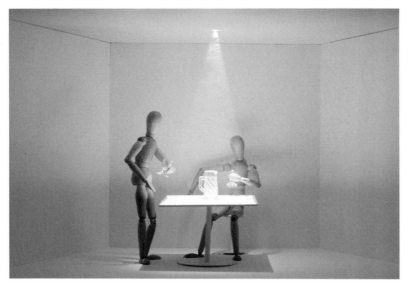

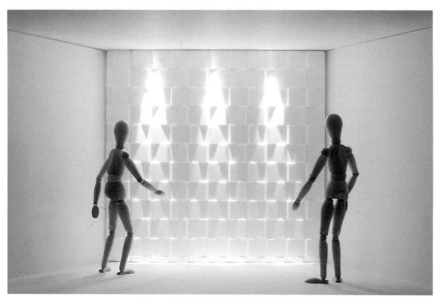

Direction, intensity and perceptual mood
The direction of a light source, and changes in its intensity, can also alter the perceptual mood of a space. A single downlighter in the centre of a space produces a visible cone of light (top left) with shadow around it. Adding further downlighters creates an ambient light, with less defined shadows (top). Directing the downlighting to wash over the end wall creates a stronger field of light in the background, weaker in the foreground (left). This effect can be emphasized using narrower beams of light (facing, top left). Washing the light over the wall (facing, top right) softens the effect. Finally, using narrow beams onto walls and as downlighter creates a dramatic sense of space (facing, below). In many cases, as here, the texture and finish of the surfaces plays an important role in the final effect.

WHAT IS LIGHT?

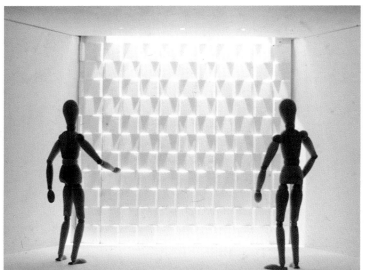

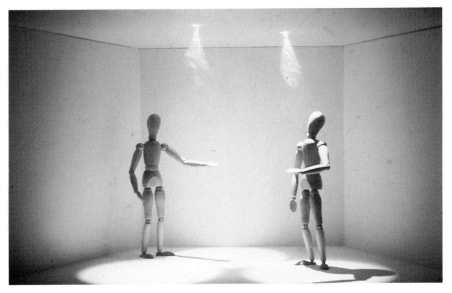

The well-known figure ground effect is one example of the way in which the brain can be confused by apparently distinct planes: is the image two faces, or a candlestick? The Dutch artist Escher's drawings also exploit the way we read two dimensional images into three dimensions (below).

been argued that the development of colour vision was of evolutionary benefit as humankind moved into an upright stance and life as a hunter.

The excellent machine of the eye is, however, not perfect. An over-abundance of light can cause an overload: the effect is called glare, and it can range from the uncomfortable, when a person 'sits with the light in their eyes' to the dangerous, when sudden light can blind a driver at night. Overdosing the eye with a particular colour can cause distorting afterimages. A positive afterimage is when the eye still retains the same shape and colour once the stimulus is removed, the commoner negative afterimage when the complementary colour is seen, as when a green dot appears in front of the eyes after staring too long at a red dot. Other more subtle perceptive tricks involve ambiguity: the well-known candlestick or two faces image, or interference images, in which the context of a shape distorts our perception of its size or form. This is linked to the phenomenon of figure-ground perception, by which the eye and brain always assume that any area separated off in an image must be on a different plane: one part must be foreground and the other background. A related effect is size constancy: because we know our vision is binocular, there is an automatic assumption that relatively smaller objects are further away.

The lighting designer needs to have some awareness of the physiological effects of light and colour, and of the visual illusions that can be created, or which must be avoided, since in most cases where ambiguity is possible in our perception of a space, lighting can be used to avoid or heighten it, as required by the brief. Carefully lit mirrors

WHAT IS LIGHT?

can give an impression of spaciousness in a restaurant, for example, but the same trick would be out of place in a workshop. Some of the perception problems we have mentioned are as much psychological as physiological.

The psychological effects of colour are well-known, even if the details and the mechanisms by which colour affects mood and emotion is not so clear. Red and orange, for example, are supposed to stimulate hunger, which is why they are widely used on the frontages of food shops. Green is supposed to be a calming or natural colour, and so on. A refinement of this theory groups colours by personality types, and by season: someone who liked browns, golds and dark reds could be said to have a calm, measured personality - an autumn person, while someone who preferred white and silver and was dynamic and extrovert would have a winter personality. There are a number of different theories, and a rather larger number of books, a few of which will be found in the reading list.

There are two important general points for the lighting designer, however. The first is the value of knowing about the effect light has on colour: how different light sources give out differently coloured light, and about how these will change perceptions of coloured objects. The second point is to bear in mind that colour is as much a cultural phenomenon as a psychological one. A simple example is the way the adjective "green" has come to refer to ecological and environmental issues as well as the colour green. More broadly speaking, colour definitions are cultural definitions. We understand the colour red to signify danger, perhaps, according to psychologists, because it is the

The same colour elements can have very different cultural connotations: red and black is the basic material of warning signs such as the familiar no-smoking sign, or the customs sign (left) while it was also used extensively by Russian Constructivist artists to suggest the power of proletarian activity.

The effect of angled lighting

These images show the different effects of lighting from below (uplighting, above), from above, downlighting (below) and with a spotlight (right). On the facing page the effects of spot (left) and floodlighting on the same subject can be compared.

colour of blood. Whatever the origin of the use of red for danger, the symbolism is now reinforced by its continual use as a danger colour. Operating theatres in hospitals were painted white to increase their illuminance, in other words for purely practical reasons. But we now speak of "clinical" white colouring, even in non-medical contexts. Different colour sets mean different things to different groups. Red, white and black, particularly used as sharply-edged solids, suggest warning signs to some eyes, Russian Constructivism to others.

Language creates similar barriers: in Welsh, for example, the same word is used for the colour blue and the colour of grass. The English word "marmalade", to describe the colour of a favourite cat, has no single word translation into French or German. Add to such problems the fact that for most people all descriptions of colour are subjective: my idea of when blue-green becomes more green than blue will probably differ from yours, for example. The lighting designer needs to be aware of such nuances, and work with the client and the brief to avoid misunderstandings and ambiguities.

The poet and author Goethe, who died in 1839, was interested in the nature of light, and performed many experiments (for all that he detested Newton's view of physics.) He proposed a holistic view of colour and light, opposed to the mechanical and divisive approach of Newton and others. Splitting light up into the spectrum told people what light was made of, but not what light was. That was something that had to be felt, as well as understood. Goethe insisted on the individual experience of colour, to the point of almost asserting that colour does not exist unless it is perceived. Most of Goethe's

theories have turned out to be incorrect (just as Newtonian physics has been challenged, though not wholly displaced, by relativity theory), but Goethe's appreciation of the subjective and psychological aspects of colour and light, and of the serious physical aspects of light and colour, is an important duality: "all theory", Goethe wrote, "is grey: but the golden tree of actual life springs up ever green!" In working with light, the potential and the reality, the physics and the experience, should always be considered.

WHAT IS LIGHT?

The simplest explanation of "what is lighting" is that lighting is the deliberate or controlled use of light. I put it this way round, rather that stressing the application of artificial light, because natural forms of light, (such as sunlight or moonlight) must always be part of a designer's lighting plan. The situations in which there is no daylight element to consider are either very deliberate (as in the theatre or cinema) or fairly unusual. Most building regulations lay down rules about access to daylight for homes and workspaces, and it is as well to be aware of these requirements when appropriate. As we shall see, lighting internal corridors is one frequent exception, an area where natural light does not penetrate and so is not an option for the designer.

Natural light

Natural light is supplied principally by the sun, and to a lesser extent by the moon and stars. The importance of sunlight or daylight is not simply that it is available and free, but that for many people daylight provides an unconscious measure of correct levels of lighting, and of light colours. This may seem surprising: after all, daylight changes in intensity and colour through the daily cycle, and can be changed further by weather conditions, particularly cloud and rain, as well as changing with the seasons. Any definition of daylight in fact has to be subject to conditions about the time of day and the weather: and 'a sunny day in June' has a very different quality of light to 'a sunny day in January'. Despite this, daylight remains an unconscious constant.

We can see this in the way colours are perceived under difficult lighting conditions.

Testing and checking lamps.

For example ,during the course of the day the colour pattern of daylight - the range of wavelengths it comprises - varies considerably. Strictly speaking, we should therefore see objects change in colour, or rather no longer recognize the same colours, as the pattern of wavelengths reflected back to us would change. This is partly true, but in addition the brain does not seem only to receive and analyse optical data, but compares them to a 'colour chart in the mind', and allows for variations in light levels in identifying colours correctly. This compensatory mechanism would appear to be innate, the result of humankind's centuries long familiarity with the cycles of day and year. It is important for the lighting designer to be aware of the contribution daylight can make to a total project, and how to take maximum advantage of daylight as needed. The colour

Light from the sky (right) underlies our innate perception of light. At the church at Ronchamps in France Le Corbusier used natural light to great effect (below right), while, In a modern office, daylight is used both from the windows on the left and through a light-trap above the right-hand wall (below).

quality of daylight is also an important consideration, especially in countries with which the designer is not familiar. Scandinavian designers, for example, are particularly adept at getting the maximum from the cool Northern light, and I personally find the quality of light in Tokyo at certain times of day very special, quite unlike the light in European cities.

A second source of natural light, flame, either in the form of candlelight or firelight, can be very useful to the designer in the right context. Candlelight has a colour temperature of about 2,000 degrees, and so produces a warm, rich light. The associations of such lighting with enclosed, intimate places, be it the corner inglenook or the church shrine, are culturally very strong, at the opposite end of an emotional spectrum from, say, moonlight, which is in turn at the other end of the colour spectrum, with a blue appearance. The cultural perception of colours, which as we mentioned in the previous chapter, is generally very important, becomes even more relevant where natural light sources are used.

Starlight and moonlight can be distant and remote: the traditional image of oak beams and candlelight (left) is much more welcoming. For a dance in the historic setting of Somerset House, the designer Paul Dyson sought to imitate the warmth and colour of fire and candle-light with pillars of light (below). These had to be built on individual pedestals, so as not to damage the fabric of the building.

These diagrams show the distribution of colour temperature for a variety of different lamps.

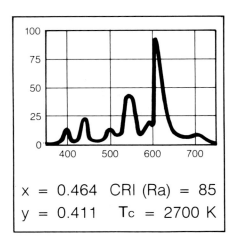

x = 0.464 CRI (Ra) = 85
y = 0.411   Tc = 2700 K

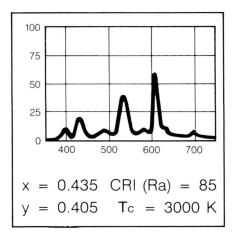

x = 0.435 CRI (Ra) = 85
y = 0.405   Tc = 3000 K

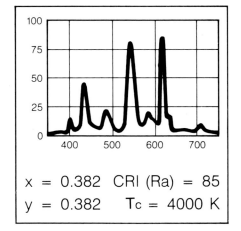

x = 0.382 CRI (Ra) = 85
y = 0.382   Tc = 4000 K

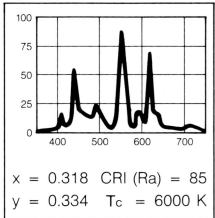

x = 0.318 CRI (Ra) = 85
y = 0.334   Tc = 6000 K

Colour temperature

One way to quantify different 'colours of daylight', and to compare these with other light sources, is through their so-called colour temperature. This is a theoretical value, based on a calculation of the temperature at which a full radiator - an ideal surface that when heated would emit one hundred percent radiation - would be a matching colour. On this basis a cloudless summer sky would have a temperature of 10,000 degrees Kelvin, a cloudy sky around 6,000 degrees, and afternoon sunlight about 4,000.

These are comparisons of intensity and colour, and another way to compare the output of different light sources is through their spectral composition. In this the proportions of different wavelengths making up the white light can be compared, by plotting them as a line on a graph. Low temperature light sources have a preponderance towards the lower, red, end of the spectrum, and high temperature ones to the blue end. This matches experience, in that we see light in a sunny sky as having a bluish cast to it, but light from an early evening sky as appearing rather warmer, due to the refraction of sunlight in the atmosphere.

Artificial light

The majority of light sources are of course artificial, using electricity to generate light in various ways. The following list gives a general specification of each type of light-source, its colour temperature and range, and any particular advantages or drawbacks it may possess. Such an outline account is obviously only a preliminary introduction, and we shall look at applications in more detail later. It should be borne in mind from the start that any lighting design is a system, that is to say it combines different elements, light sources, power levels, surfaces and colours into a total whole. It is achieving the whole effect that is the task of the lighting designer, and the whole solution must be thought of as a system, not a collection of disparate elements, if the design is to succeed.

There are two broad classification of electric light-sources, incandescent and discharge, terms which relate broadly to the operating system of each. Incandescent lamps were the first to be manufactured (the honours of invention in the late 1870s are shared by Thomas Edison in the USA and Joseph Swan in Britain), and so should be dealt with first. The incandescent lamp - what we term the ordinary light bulb - generates light by passing a current through a wire coil, mounted in a vacuum (or in an inert gas) until it is incandescent. Early lamps used a vacuum only, and later iodine or bromine was added to the vacuum to improve the performance of the lamp.

Discharge lamps work by passing electric current through gases or metallic vapours, again contained in a bulb (mercury vapour is one of the commonest used). This

Three halogen mains voltage lamps: Par 20, 30, 38.

Different lamps colour the same white object in different ways a metal halide lamp is bluish, a white son lamp slight yellowish, compared to a tungsten halogen (reading left to right).

(Facing page) With a coloured object the changes in apparent colour due to different lamp types become more complex: top row, left, tungsten halogen; right, metal halide; bottom row left, white son; right, tungsten.

produces a fluorescent discharge, visible as light when it passes through the phosphor coating on the inside of the bulb or tube. At high pressures such lamps are efficient, a good part of the radiation produced by the discharge being in the form of visible light (incandescent lights, for example, produce too much energy in the form of heat, which is inefficient and can cause design problems). The sodium lamps mentioned above, and the mercury halide lamp, are well-known types of these lamps. Fluorescent lamps, often called fluorescent tubes, use the same discharge effect, but at lower pressures: they offer a range of different gas mixtures within the bulb, producing a range of colour rendering effects depending on the precise phosphor coatings used.

Taking the incandescent lamp as a starting point again, we can compare its colour temperature against its efficiency and against its ability to render colour, and make the same comparison with and between other lamp types. An incandescent lamp has a colour temperature of about 2,750 degrees, and as one would expect from such a low temperature the colour definition of reds and oranges is strong, but the blues are dulled. The efficiency of the lamp, that is to say the number of lumens generated for watt of electricity used, is also quite low, being about 15 on average. So a 75 watt lamp will produce about 1,000 lumens. Because we are so familiar with this type of lighting, at least in a domestic context, we see this level of lighting and colour rendition as acceptable. The improved tungsten halogen incandescent lamp has a colour temperature ten percent higher, and so better rendition of blues, while it is also more efficient, giving on average twenty lumens to the watt. A 75 watt

WHAT IS LIGHTING?

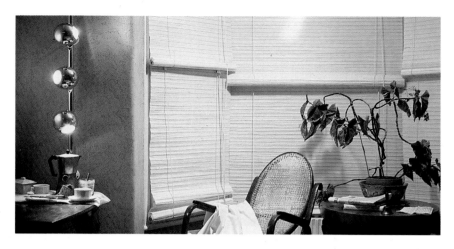

Mains track systems can accept plug-in low voltage spots with integrated transformers (above), or dedicated low voltage track can be used (right). The detail (below) shows the interior of a track system with the two sets of conductors and the probe contacts.

High and low voltage systems

The two most important areas of development in the recent history of lighting are in the introduction of low voltage systems, and of new lamp types both for high and low voltage. I sometimes think that the array of new and existing lamps available creates too much choice for the designer! The reasons behind the creation of new lamps are various. In some cases a new engineering process makes a combination of filament and gas economically viable for the first time. In others the desire to create lamps with good environmental characteristics has been important. The main considerations are the light output ratio, lamp life, consumption efficiency and the quality of light.

Each new lamp invites those of us who design fittings also to think again. How best can this lamp be put to use? Can we adapt or modify an existing fitting, or do we need to start from scratch with a new approach. Many modern fittings are designed to support various different lamps. This is economically efficient in production terms, but more importantly allows the end user to mix different light sources into a coherent design.

The option of main or low voltage systems has also enlarged the choices for the designer. Low voltage systems were introduced for several reasons. Firstly, low voltage lamps, and therefore fittings, were considerably smaller. They could be placed more discreetly. Secondly, though the capital costs were higher (low voltage lights require transformers to step down mains voltage to low voltage) they were much more economical. Thirdly, because of their lower wattage, they generated less heat at the light source. Fourthly, the smaller light sources allowed for more precise reflector design than had been possible before.

Another development pushed low voltage forward. Transformers had been traditionally bulky and heavy objects, requiring ventilation. A new generation of small electronic transformers, that weighed much less and produced less heat, became available. Instead of placing the transformer behind a bulkhead or within a ceiling space to keep it from view, the transformer could be made part of the fitting, a discrete box integrated into the overall shape. This in turn brought another advantage. High and low voltage lights could be attached to the same track,.

Technology was also coming to the aid of the fluorescent lamp. Long regarded as the poor cousin of discharge lamps, fluorescents had been banished to factory floors and other purely utilitarian spaces. One problem was the strike-up time, the delay between switching the lamp on and it's becoming fully lit. This could be several seconds. By developing electronic ballasts that controlled the start-up procedure and regulated the voltage flow to the lamps, fluorescent lamps became as prompt as incandescent ones, and the balanced flow of current increased lamp life. Better colour quality and low energy requirements also encouraged the development of compact fluorescents.

This development also allows for a new range of fittings, and the fluorescent lamp is again an important part of the lighting armoury. The most recent development is the miniature fluorescent tube, with a diameter similar to that of a cigarette, but producing as strong a light as a conventional lamp. These new lamps will offer more choice, and more challenges, to designers.

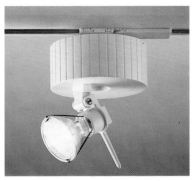

The different approaches to transformers: Tiller spotlights with a transformer in the ceiling (above, left) or integral unit (above, right). Electronic transformers (left) are much less cumbersome. Low voltage aids miniaturisation: the scale of some new fittings is remarkable (below), as is the size of the new FM miniature fluorescent (bottom).

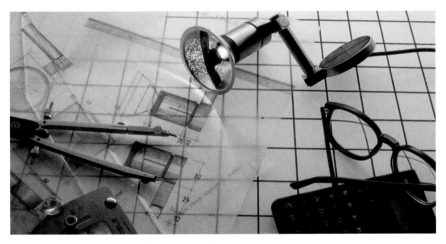

Fibre-optic lighting

Another new development with great potential is fibre-optic lighting. This relies on a well-known effect, whereby light passing down the inside of a column of glass is totally internally reflected. So the light does not escape through the walls of the column, but only through the plain face at the end. The effect works whether the column is straight or curved. As glass columns are not very practical as lighting units, the effect remained a curiosity until the development of fibreglass, which also exhibits the same properties.

Individual strands of fibreglass can be bunched together to create long, flexible light diffusers. A light source at one end and a diffuser at the other are all that is needed.

The advantages of such a lighting system are several. Firstly, since the lamp is at the remote end of the system, the light source is a cool one. Secondly, the divisibility of the fibre-optic source makes it very versatile (one fibre-optic table lamp consists of several hundred strands, each a single tiny source of light, that float free above a fixed base, like the seeds on a dandelion.) Thirdly, the flexibility of fibre-optic conduits make them useful in contexts where there is no space for a conventional bulb. Finally, because relamping is done at the source, fibre-optics can be used to advantage in settings where security is important.)

One obvious use for these lights is in display (especially in museums where security and climatic control are important. Another use is to provide background lighting, giving a ***pointilliste*** effect over a large area. Other uses will certainly appear as this new technology develops further.

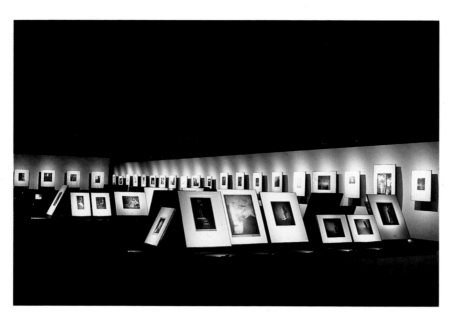

Fibre optic lamps in use for at exhibition of early photographs at the Musée Rodin, Paris. The cool light they deliver is very important where the conservation of works of art is an issue. Fibre-optics also provide safe light, as in this swimming pool (right).

Lights as lamps and fittings

The full range of all available lights is immense, and we do not propose to catalogue it here. For our purposes, here is a shortened list of the main kinds of lighting fixtures available to designers, preceded by a diagrammatic list of the range of different lamps available. The pairing of fitting and lamp (as many fittings accept a range of lamps) is the detail in a lighting design; the planning of the design as a total system, and not just a set of parts, is the central element.

The gallery of lamps (on page 141) shows, for each model, the shape, available wattages, output, lamp life and international designation, with notes on the group of lamps it falls into, and the colour rendition it offers. But note that colour rendition and colour temperature varies not only between the type of lamp but between different manufacturers, and the values given here, while correct for Sylvania lamps, may not be exact for all other makes. If the colour temperature or another variable is critical, it is best to go to the actual manufacturer's handbook.

Turning to the range of fittings, the designer has a basic selection of types to choose from: spotlights, floodlights, downlighters, uplighters, wall-washers, general luminaires and fluorescents. With these there is also a range of mountings and accessories - track or wall-mounted, recessed or exposed, with filters or barndoors, single or multi-circuit, with dimmers or control systems. And for many fittings, except fluorescents, there will be an additional choice between low voltage and standard voltage versions. Let us look at each of these in turn.

These four installations (including Bentalls store, top left) show the potential of fibre optics for decorative lighting.

Spotlights

Spotlights are intended to give a narrow beam of light (typically 5 to 30 degrees beam angle) from a single source. As spotlights are regularly used to highlight features (for example in a retail display) they are normally provided with tilt and swivel facilities, and sometimes a degree of focusing. It is also important to check that there is no backwash of light from the end of the fitting. The choice and form of the reflector (either integral with the lamp or part of the fitting) will affect the intensity, sharpness of edge and colour of the light emitted by a spotlight, as will the choice of lamp.

The Talus spotlight is asimple mains voltage range which will take a range of lamps. It can be used with the Par 30 lamp, widely available in wattages from 60 to 120, as well as the new 100 watt Par 38 lamp. It is also available for a crown silvered 100 watt lamp: the silvering eliminates direct glare from the light source. The housing can rotate through 345 degrees and tilt through 90 degrees. The light can either be mounted directly on a wall or ceiling surface, or onto a track system.

In the low voltage range, the Infinite spotlight offers a similar versatility. It can be fitted either with a 20 watt spotlight, with alternative beam angles of 13 or 31 degrees, or as a 50 watt dichroic spotlight, or as 50 or 75 watt capsule spotlight - again with a choice of beam angle of 7 and 30 degrees at 50 watts, or 7 and 40 degrees at 75 watts. An adjustable 13 to 19 degree fitting is also available in 75 watts. The light can be mounted individually or onto a track system, and a projecting handle allows for rotation and tilt when the lamp is lit.

The municipal arts centre at Soest, in Germany, lit by Talus spotlights (above). Talus with Par 38, Par 30 and parabolic fittings (below, left to right).

Infinite spotlights on a track mounting used to illuminate a recent Design Council exhibition, (left) with different lamps: 75 watt capsule spot, (top) and 50 watt dichroic spot (above).

Floodlights

Floodlights project a wider beam of light. Unlike a spotlight, where the beam can often be focused, a floodlight projects a field of even light. But as with spotlights, floodlights are intended to provide a general lighting accent, and so often need to be fitted with tilt and swivel facilities. To limit the spread of light barndoors can be fitted to floodlights (as they can to spotlights as well) to define edges clearly, and discourage glare. It is also important with floodlights to ensure that the light distribution is even over the whole area lit, whereas with a spotlight variable intensities of light are acceptable, or can be controlled within a narrow beam by iris units or Fresnel lenses.

The Powerflood, for example, is a compact floodlight using 200 to 300 watt linear tungsten halogen lamps, which can be fitted with either extension hoods or barndoors to modify the spread and field of the beam. Often spotlights and floodlights are to be used together (for example on the same track) and so need to be designed to look appropriate when adjacent. One answer is to design a basic fitting to which different lamps can be attached, with additional features such as cowls, to reduce side and backlighting, or parabolic reflectors.

Top of the range systems, such as the Control Spot series, blur the distinction between spot and flood completely, in presenting a series of compatible and often interchangeable designs that offer an array of lighting solutions in a modular form. The Projector accepts fittings for a 50-100 watt low voltage tungsten halogen lamp or the warmer 100 watt white son lamp. The low

Powerfloods with cowls used as uplighters in an art gallery (above), with the alternative fittings of cowls (below) or barn doors (bottom). 300 watt linear tungsten halogen lamps are used in each case.

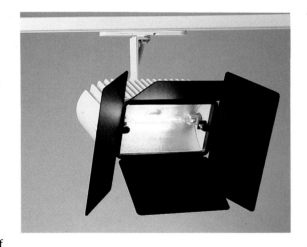

WHAT IS LIGHTING?

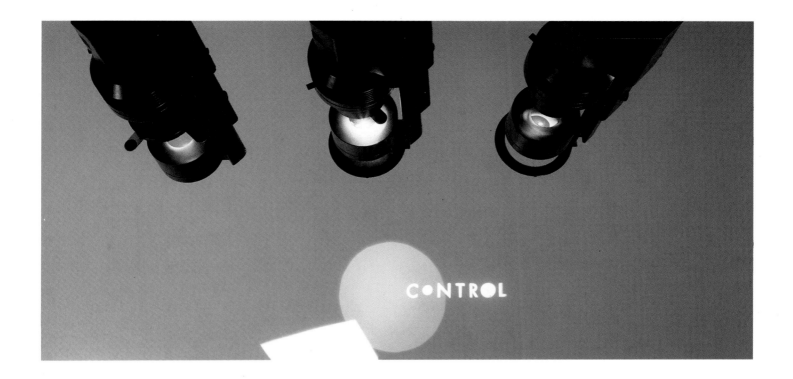

CONTROL

voltage lamp comes with three possible beam angles, the son lamp with two. In either case the lens position can be adjusted to give a soft or hard edged beam. The same basic family also includes low voltage and son spotlights, a metal halide spotlight, a par 30 and par 38 spotlight, and an a-line halogen spotlight. Across the whole range filter holders for coloured gels and U.V. are available, and for the projectors a range of lenses for focusing narrow to wide beams are available, as are iris heads, framing heads and gobo holders to generate a variety of effects. On the spotlights, there are variable reflectors, glare shields and barndoors also available. The range of lighting choice, within a consistent design idiom, that this kind of product offers, is extremely important for the modern lighting designer.

Control Spot projectors, with framing head, iris head and graphic gobo projectors respectively (top). The range of fittings for the low voltage spot is below: alternative reflectors, glare cowl, barn doors, U.V. filter and gel filter.

WHAT IS LIGHTING?

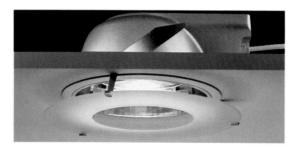

Four different downlighters: the internal reflector, though out of view, has an important influence on the final light effect - (Equinox (above), Solstice (top) Par 30 (middle) and LED 100 with floating glass ring.

Downlighters

A downlighter, as its name implies, directs light downwards, either as a narrow or wide beam, depending on the intended effect. Downlighters are often recessed wholly or partially into ceilings to create concealed lighting effects and to reduce glare. As with other types of lighting, a variety of lamps can be supported by downlighters.

One modern system is the Advanced Equinox. This is a mains voltage high output light, available with 100 watt white SON, or with 70 or 150 watt metal halide and tungsten halogen lamps. With the metal halide lamps, different beam widths are also available, from 25 to 70 degrees. Pinholes, U.V. filters and spillrings (to reduce glare) are available, as are floating glass suspended discs. These accessories for downlighters soften the light effect by directing light back across the ceiling, and reducing the visible edge of the fitting. Colour tints in them can also change the light colour, though not so dramatically or accurately as a colour filter.

A typical low energy light is the Solstice. Like Equinox, it is designed for a recessed ceiling housing. It is available with a range of lynx lamps, of either 5, 7, 10 or 13 watts, and with a range of external accessories, including glare controls and diffusers. These fittings can be used independently or together, and an emergency version is also available. This may seem a simple list, but it gives a total of over 40 variations on the basic fitting, and a similar breadth of choice to the designer. Recessed downlighters of the eyeball type can also permit a certain degree of angling, thus providing a downward spot effect. The Chorus series is an example: it has a choice of four for different beam angles.

Uplighters

Uplighters project light upwards, normally onto a ceiling to create an overall indirect wash of light. For that reason the beam angle is important, and most uplighters are available in a wide choice of beam and with a range of lamps, to exploit different colour effects. Clearly, with most of the illumination being created by reflection off the ceiling, the ceiling finish and surface material is particularly important in deciding when to use uplighters and where to place them. Uplighters can either be freestanding or suspended from a ceiling, as well as corner- or wall-mounted.

The Quill uplighter, for example, accepts double-ended tungsten halogen lamps up to 500 watts, or 70 to 250 watt high intensity discharge lamps. The same housing design is available in a freestanding form, as a unit suspended from the ceiling. These models have a full circular cone. A semicircular wall-mounted unit is also available, as is a quarter circle corner unit. The surface-mounted models have a variable blind fitted to adjust the light distribution, while the freestanding version can have symmetrical or asymmetrical reflectors fitted to move the light course slightly off centre. The finish of Quill uplighters is plain. As the essence of uplighting is in letting the ceiling do the work, most uplighters are deliberately discreet in their surface design and visual qualities, unlike spots or floodlights which may need to signal visually their presence as light sources.

Freestanding MIL uplighters (left), wallmounted Quill (below)

Uplighting

The versatility of uplighting can be seen from its use in these very different settings: uplighters highlighting a ceiling (top), wallmounted for a domestic living room (below, right) or supported by daylight in an office (below, left).

Downlighting

Downlighting is used effectively in this rather grand entrance (above) where it also draws the user's attention to the steps. Downlighting can add drama to interior halls (below, left) and office spaces, especially if narrow beams are used (below, right).

Wallwashers can be mounted, as with Mura (below, right) or recessed (Equinox, right). Note the angling of the Equinox housing.

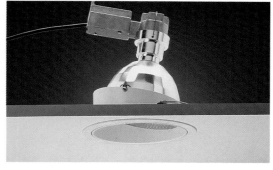

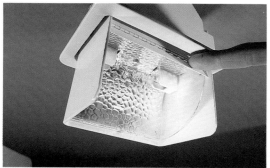

Wall-washers

Just as the uplighter projects onto the ceiling to obtain a diffused light effect, so wall-washers project light over a wall surface. As with uplighters, much of the effect of a wall-washer will depend on the material and finish of the wall, and the light fitting itself is usually a discreet one. Normally light is projected from the top downwards, but this can of course be reversed. Very often the fitting to provide wall-washing light will be a downlighter with an adjustable opening or filter to direct light against the wall: the Advanced Equinox, for example, is available with just such a kick deflector.

The Mura wall-washer is available either as semi-recessed or ceiling mounted, to direct light down the wall, or with a floor mounting plate to direct light up the wall. The fitting takes 100 to 150 watt tungsten halogen lamps.

General Lighting

Old fluorescents did not provide much visual stimulus (below). New fluorescent lamps are more compact and efficient (right).

Although modern fluorescent lighting can match some of the lighting conditions created by incandescent lamps, it is best to treat fluorescent lighting separately for two reasons. Firstly, the size of the fluorescent lamp is often radically different, and secondly fluorescents only provide general illumination: they cannot be used as spotlights, for example.

The simplest fluorescent fitting is a lamp and holder, often ceiling mounted. This can be refined by adding a diffuser below the lamp to reduce glare and soften or even tint the light. In an independent fluorescent unit, the key element may be the louvre and diffuser to be fitted, since this choice will

WHAT IS LIGHTING?

have an important effect on the overall light. Further choices are between compact and standard fluorescent luminaires, and whether to ceiling-mount, recess or suspend the fittings. A suspended unit can then be linked into a system of multiple fluorescent lights. The suspended unit can be designed to direct and organize the light, for example by providing a clear top to the unit, so allowing it to act as an uplighter or as an uplighter and downlighter.

The colour potential of fluorescents is well illustrated in these two examples, above, both of offices. For a staircase, (above left) fluorescent tubes have been fitted under the glass treads, while in the corridor (left) fluorescents are used together with downlighters and daylight.

Linear systems

Linear systems may be made up of straight elements, but need not be dull: here they curve or snake down corridors (right, above and below).

Very often the lighting designer will need to link a series of fluorescents together to achieve a total scheme, and this is where linear systems come in. They allow for joining fluorescents together in matching housings, with corner and straight joints, so as to meet the requirements of varying floor plans.

Linear systems such as Fluorotube and Lytetube can incorporate a variety of different lamps, in differing combinations: single or double standard fluorescents, ceiling or suspended mounting, are all possibilities. The orientation of the housings can be adjusted to provide downlighting, uplighting or a side beam, and lighting units can be replaced with track systems (see next section) to support spotlights.

Ambit is another such sophisticated system. Its trapeziform section allows for two light fittings on one side, one on the other. These can be all fluorescent, providing two downlights and one uplight (or vice-versa if the unit is turned over), or a mixture of fluorescent and incandescents, including spot and flood lights, wall-washers, and so on. The individual units, typically 1.5 metres long, can be joined directly, at right angles, into T or X junctions and across a curve. This versatility of positioning and fittings (and many fluorescent systems are equally flexible) makes fluorescent lighting an important tool for the lighting designer. It is also particularly important with fluorescents to check on the lamp colour. Fluorescent lighting was often considered harsh and inelegant: modern fittings and lamps show this is no longer so.

Linear systems can be articulated in the horizontal and vertical plane, making them able to follow a staircase (left, at Evesham Library). Even an open square grid can provide visual interest (as in this canteen , below right). Straight fluorescents can be used to good effect when made a regular feature, as in the office (below) where they space out the ceiling joists.

The Queen Elizabeth II Conference Centre in London, where a sophisticated control systems allows for different events: lectures, conferences, presentations and performances.

Dimmers, multiple circuits and control systems

A dimmer is a simple device that by reducing power reduces light output. It can be used to change the mood of a space, or to balance different levels of daylight and artificial light throughout the day. A multiple circuit track system allows for alternative lighting solutions, depending on which luminaires are connected to which track. A control system combines both elements, dimmers and selections of lighting solutions, into a single package.

Control systems can be set up so as to command and control several different lighting circuits and dimmers. These can be operated individually, or selected arrangements can be programmed as "scenes", or complete alternative solutions. The key to a successful control system is that the user should understand it, while it should, ideally, provide for as many alternatives as possible. The lighting designer should bear in mind that few spaces remain immutable - new equipment is fitted in offices, new exhibits arrive in museums, new tables in restaurants and new furniture in homes! A good lighting design will meet the client's present needs and anticipate future changes, by providing flexibility in fittings and control. Studies have also shown that giving the user a degree of control over the lighting of individual work spaces has important benefits in terms of efficiency and employee satisfaction, and the designer should therefore think of the end-user, particularly in office designs.

WHAT IS LIGHTING?

Two views of the same boardroom, showing two different lighting scenarios in use: with centre down-lighters in both cases, and with uplighters on the end wall (left).

Track systems

Track systems offer a wide range of options to the lighting designer, and allow the end user the opportunity of modifying a lighting plan without extensive refitting. The underlying principle is simple: like a railway line, the track allows the light fitting to be fixed at any point on its length, and different types of fitting - uplighters, spot or floodlights - to be mixed. Corner and crossover units allow a track to be deployed in almost any space.

Tracks can either be ceiling mounted or suspended, and can be fitted with single or multiple circuits, allowing for different lighting options. Because the position of individual luminaires can be adjusted endlessly, tracks are particularly useful in situations with changing lighting requirements, for example in shops, where new displays replace old, or in offices where desk plans are altered.

The Infinite low voltage track is designed for use with the Infinite luminaires described earlier. It can be suspended or ceiling mounted, and the luminaires either fixed directly into the track or further suspended below it. One interesting departure is the availability of a curved section of track, which when used with a variable coupling device allows for a much wider range of applications than a traditional rectilinear system.

Lytespan 4 is a sophisticated mains voltage system that accepts four circuits, into which any of the luminaires in the Lytespot range can be fitted. As with most track systems, illuminated signage can also be incorporated into the layout, an important consideration in public access spaces. A low voltage version of Lytespan is also available.

The Infinite track system in a domestic setting, showing how it can be adapted to the demands of a complex space.

General luminaires

The designer should also be aware of the range of individual general fittings on the market, from simple incandescent bowls for ceilings to lights for gardens or exteriors. The fact that these fittings are straightforward does not mean they should be overlooked! As for the sophisticated "designer lamps" often created by architects or industrial designers, too many of these "light objects" can clutter up a design, but they can be used to add an occasional accent to a balanced design. A successful lighting design should be efficient and unobtrusive, however.

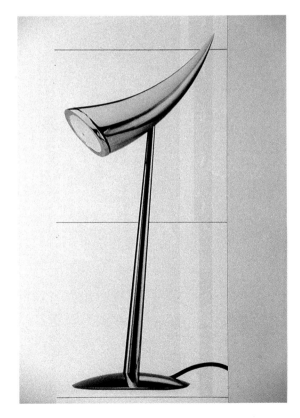

Designer luminaires can be dated back to Charles Rennie Mackintosh's elegant Art Nouveau curves (above) or come up to date with Philippe Starck's cheeky Ara lamp (left): the most tactile of task light.

This antique sculpture has been lit using a Control Spot: note the angled box deliberately produced on the framing head to create an unusual effect.

LIGHTING IN USE

The lighting solutions possible for any task can be divided into a number of categories. These categories do not represent solutions in themselves, of course, but approaches to a total solution. They can be seen as paired opposites, as follows: daylight and artificial light, direct and indirect lighting, uplighting and downlighting, wall-washing and spot-lighting. This section will look at applications of these categories.

Daylight and artificial light

As we saw when looking at specific projects, one of the first questions the designer must ask is about the level and quality of daylight in the space to be lit, and its sources (for example, whether it is direct sunlight or reflected off a facing building or through a garden) as these sources will change the quality of the light, as will the orientation of the room with regard to the sun's path in the sky. The hours in the day the room is to be normally occupied will also affect this calculation.

A further question for the designer is whether there are any specific obligations in providing light levels to be met (other than any set out in the brief). For example, building regulations often require spaces to which the public have access to be lit to minimum levels. If the scheme to be lit is a complex one, with several spaces included in it, the changes in light level between different areas will also need to be taken into account. Such changes can be key elements in making a brief work, for example in moving from a busy hotel lobby into the more relaxed surroundings of a restaurant, or from the public space of a showroom into the more private space of an office.

This serene anteroom benefits from combining daylight diffused through the glass door surrounds and cool metal halide uplighting on the walls.

LIGHTING IN USE

Daylight is free, and it's friendly. It should be the designer's starting point, either opening up the space to daylight, as in the Euroclub Lounge at Heathrow Airport, (above, designed by XMP), where the large curved glass roof adds to the sense of space, although supplementary lighting from tungsten halogen downlighters, will be needed for spaces outside direct daylight and for evening use (below). Daylight can be used, more indirectly, through a light well as in the Neal Street Restaurant in London (facing page, top), where daylight, painted brickwork and foliage combine to give a Mediterranean feel. Recessed downlighters in the well can maintain the effect at all hours.

As daylight comes and goes, the lighting designer needs to allow for this. In planning the lighting of the Isaac Newton Institute building at Cambridge, care was taken to let the evening lighting effects (below right) have the same balance as the daylight ones (below left).

Textures

The way light falls on a textured surface is important: the draped cloths in these two illustrations are very similar but lighting them from behind (facing page, right) creates a softened veil of light, while light falling directly (facing page, left) creates a solid patch of illumination. Texture can also be lit to create atmosphere: in this cinema in Courtrai, Belgium, the red light from Par 38 lamps and the red wallcovering provide a warm welcome (below). At the Hayward Gallery discreet downlighting was all that was needed to bring out the textures and reliefs of Mark Boyle's work (above, right). These rubber tube inner sculptures (above left) by Susan Stockwell, were lit by grazing light.

Texture, reflection and transparency

After the natural light available in a space has been considered, the next question is the materials from which the space is made. Textures have a practical consequence, as we have seen earlier, in that they absorb and reflect light at different rates for different surfaces. Over and above the practical aspects, the aesthetic effects of textures in diffusing and transforming light are significant.

The predominance of a particular surface colour will also have its effect on the colour of overall light in a room. (The French National Railways use a standard strip light mounted above the windows in their

Reflections

Using mirrors to increase the apparent size of a space is a familiar device, seen here in a cool and clear living room (above left) and with the added effect of reflective pillars in a banking hall (Seaman's Bank, New York, above.) Variations on this theme include using dramatic light fittings and furniture to highlight the mirror itself (Katherine Hamnett shop, facing top) or hiding the edges with floral displays, as in the dining room (facing below). Using curved walls to distort the effect can also be effective (facing, right) in a private dining room. Note how the highly-polished tabletop enhances the effect.

The varied textures of Andrew Logan's sculpture of Pegasus (left) - in metal, glass and fabric - provided an interesting challenge to the lighting designer: we used a combination of uplighting and spotlighting to dramatize the object, emphasising reflections off the mirror chip wings.

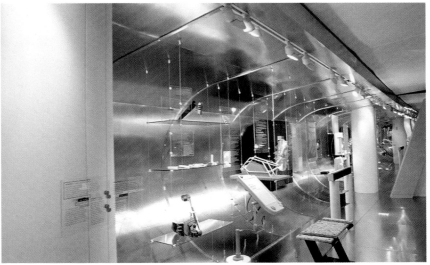

coaches, but non-smoking coaches have white ceilings, smoking ones a mid-grey. The atmosphere - no pun intended - in each is quite different, despite all the other features being the same.) Often, the colour of wall and ceiling finishes exerts more effect on the apparent size of a space than lighting can. A space fitted out in dark colours will always seem smaller than the same space in pale colours, however much lighting is added to it.

One way to increase the apparent size of a space is through reflective surfaces. This is something that needs to be used with care: it is not a solution appropriate for an office space, for example, though it could be in a restaurant. Another way to increase apparent space is through transparent surfaces, either to take advantage of a natural view, or for a more practical reason, for example in overseeing a workspace. The illustrations on these and the following pages show some of the possible interpretations of these solutions.

LIGHTING IN USE

Transparency

Because looking at things through glass always seems to make them interesting, the traditional glass showcase (facing, top left, from the Oggetti shop) can make an effective presentation if well lit, here with recessed dichroic 50 watt tungsten halogen lamps. For a radical use of transparency, see the backlit fish tanks in Katherine Hamnett's Sloane Street shop (facing, top right), designed by Nigel Coates, and for a practical application the Metropolis recording studio, designed by Julian Powell-Tuck (facing, below) where the glass wall separates and communicates performer and recording engineer. Transparency and reflection combine in Simone Kesselheit's fabric hangings (right).

A professional work area such as the advertising office of Loot magazine (right) needs an even undifferentiated light which will interfere minimally with working VDU screens. A meeting space (below) may be amenable to more variations, which can also help to isolate parts of the work area from others, as in these offices for the Banco d'Italia.

Direct and indirect lighting

Direct and indirect lighting may also be used to highlight changes of mood, atmosphere and purpose. An architect's office, for example, may need direct and even lighting on workspaces and drawing boards, while reception areas and meeting rooms would be better lit indirectly. In planning direct lighting, the positions, locations and physical attitudes of the intended users may be important, to avoid problems of glare, in particular. Many schemes not only use a combination of direct and indirect lighting but provide for switching between the two alternatives, or supplementing one with the other.

Direct lighting need not be dull: this translucent 'light-pipe' is a major visual feature in an architect's offices (left). Indirect lighting can often be dramatic: this room setting (below, left) is from a special display for Carlton Hobbs, specialists in antique furniture and objects. Control Spots were used here because of their versatility: the large deep blue disc is the light from an antique glass globe. Low-voltage spots can be equally effective, as in the restaurant of an arts centre (below, right).

Uplighting

The versatility of uplighting can be seen from its use in these very different settings: wallmounted 150 tungsten halogen for a simple office space (right) or used to highlight fine old plaster work in a historic house (facing, left: the picture shows Somerset House in London transformed by Paul Dyson into he setting for a Georgian Rout), or supported by daylight in a formal entrance hall and accenting the plasterwork and ceilings (facing, right). On the facing page the 1970s splendour of the Biba Rainbow Room restaurant was achieved by uplighting the recessed ceiling (below), using coloured gels around continuous fluorescent tubes.

Uplighting and downlighting

This is in some ways a special case of the direct/indirect distinction. An uplighter throws light upwards, to reflect off walls and ceiling (or just off the ceiling in the case of a freestanding fitting) to provide indirect light for a space. Downlighting is the reverse: independent units fitted in a ceiling (or on a suspended track) spread light directly downwards, or wallmounted fittings wash light downwards over walls to provide general illumination. The effects of such lights, especially when wall mounted, will depend very much on the nature of the wall surface,

LIGHTING IN USE

its texture and reflectivity. In a large space uplighting can be used to increase the sense of space, particularly if the ceiling design and material is interesting in itself, whereas downlighting in the same space can be used to add drama and detail, for example in lighting a church or historic building.

Downlighting

The pools of light cast at floor level by downlighting can give regularity and depth to a plain corridor (above), but can also be used to increase the complexity of a space, for example in the receding series of door frames (left) or in animating the entrance to a concert hall (facing page). Downlighting can also create deliberate pools of light at floor level, as in the gymnastic club in Germany (facing, below).

Wall-washing and spotlighting

Wall-washing means projecting light over the surface of a wall to provide indirect lighting for the adjoining space. A key problem for the designer is to place the light fittings so that they distribute light evenly over the wall, without overspill and without glare, particularly for someone near the wall. Spotlighting means highlighting an area or an object with a narrow and often intense beam of light. It is a dramatic way of drawing attention to a particular feature - for example the key exhibit in an art gallery or trade fair display.

LIGHTING IN USE

Eva Jiricna's interior for the Joseph shop in Knightsbridge, London, combines downlighting dichroic wallwashers with reflective surfaces and plain textures to create a cool, high-tech environment (facing page), while plain areas of washed light, floor to ceiling, frame the displays in a photography gallery (facing, below). Wall washers in the exhibition area at the National Gallery, Berlin (above) also concentrate attention on the walls - an effect heightened by the deliberate choice of dark carpeting in the central area. The lighting consultant Hans von Maloti worked closely with the architect Mies van der Rohe on this scheme: it was installed in the 1960s, and is as effective a design now as then.

LIGHTING IN USE

Spotlighting

The use of a spotlight to highlight a sculpture for a
Fashion Gala at the Royal College of Art creates a centre
to the room (facing, top), while the spotlit picture in a
reception area also provides a focal point, while allowing
enough general light for work (facing, below right).
Spotlights focus attention on the motorbike displayed at
the Design Museum (facing, below left). Alternatively
spotlights can highlight areas (as in the restaurant at the
Halcyon Hotel London, above right) designed by Julie
Hodges, or dramatise an object, such as the antique
mirror (above). Paired overlapping spotlights can also
enhance texture effects, as in this doorway at the Burrell
collection, Glasgow (right), designed by architect Barry
Gasson.

This restaurant/canteen (above) is divided into a number of rooms. These are separated by grey grilles in the wall openings, allowing views across the different spaces. The grey grilles and blue chairs provided the main colour notes: the walls and coffered ceilings are off-white. Downlighters were fitted between the ceiling coffers: these lit the central tables and provided incidental colour effects off the coffers themselves. Wallwashers on the blank wall spaces built up the overall light to the required level, and allowed the different spaces to interact without imposing.

Combined lighting and ambient lighting

The examples on the preceding pages have been chosen to isolate approaches to lighting, but more often than not a successful design combines several different approaches. Here the designer needs to mix the effects of light from several sources into a satisfactory whole. Tracked systems are very popular precisely because they normally support different luminaires and lamp types, and allow not only for trying out different solutions on site but also for modifying the lighting layout if the use of the space or the user's requirements change.

I often try in planning a combined approach not to choose a first solution - downlighting, for example - and then see what I need to add to it to get the space to work, but to think in terms of a dual approach from the start. It's the balance between the elements that is important. Often the space itself will help find an approach: a

hotel lobby, for example, needs to concentrate light on the entrance itself and the reception desk, to welcome the visitor in. A company boardroom will need an even level of light all around the central table, without favouring a particular part of the room, while a workspace such as a hairdresser's salon will need emphasis on the customers' chairs and a softer atmosphere elsewhere. If the space is one through which people move, then the plan of the lighting will need to encourage this movement, as in a office reception area or airport lounge. These considerations often help to formulate an initial design idea, which is then refined by further study, site visits and technical calculations.

Often the lighting designer's task is the straightforward one of providing even illumination over a working area: a classroom, a kitchen or an office, for example. This kind of ambient lighting can still be open to imaginative solutions. I find it useful in such a context to try and plan into the system some elements that can be changed and modified by the user, for example by putting part of the total lighting on a separate set of switches, or providing dimmers. Letting the client or user know that they have some control over the system, and can adapt it to their changing needs, is a very real aspect of a successful lighting design.

For this 'luggage shop' (above) the client wanted all the emphasis on the goods displayed: the wall colours are neutral as is the floor. Fitting wallwashers to light the integrated displays would have left the centre of the room dark, so this was supplemented with a row of wide-beamed downlighters in the centre of the space to light a display at floor level.

Jewellery requires good clear light to be seen to its best advantage. In lighting the showrooms of the London jewellers Garrards also wanted to preserve the nineteenth century decorative features and lamps that contributed much to the atmosphere of the showroom (right). The solution was to leave the antique ceiling pendant luminaires in place, and add a specially built linear system suspended from the ceiling housing 50 watt low voltage adjustable parabolic reflectors to spotlight the central display tables. The wallmounted display cases were also lit internally, and desk lamps provided detail lighting for examining individual pieces of jewellery.

LIGHTING IN USE

In this foyer area for Stanhope properties, the floor plan helped define the lighting arrangement. The spinnaker design in the floor was repeated in the placing of downlighters in the ceiling, while the overall level of light was supplemented by wallwashers and the ambient light through the glass bricks in the facing wall (above).

At the showroom for Nagabori, Tokyo (below) the curved display wall is lit by variable focus tracked spotlights, while the meeting tables are more discreetly lit by downlighters in the ceiling. The Woburn Abbey restaurant (below, right) bounces light off the ceiling, combined with downlighting onto the tables.

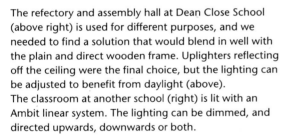

The refectory and assembly hall at Dean Close School (above right) is used for different purposes, and we needed to find a solution that would blend in well with the plain and direct wooden frame. Uplighters reflecting off the ceiling were the final choice, but the lighting can be adjusted to benefit from daylight (above).

The classroom at another school (right) is lit with an Ambit linear system. The lighting can be dimmed, and directed upwards, downwards or both.

Special lighting fixtures
displayed at the Concord
Gallery, London.

HOW TO USE LIGHT

Having looked at simple projects, I want to turn to more complex ones, to see again how the lighting designer assesses and executes the brief, and what problems can be encountered along the way.

This section looks in detail at the planning of some straightforward lighting designs. We saw earlier that a lighting design must be a total concept, not merely a collection of luminaires. The designer's job is to plan and organize that concept, in the most efficient technical way, so that it is safe (a key matter) and simple to use. To do this the designer must first listen to the client's needs and requirements, and then study the characteristics and problems of the space to be lit, and then find the right means of presenting design ideas.

Plans and elevations

The lighting designer traditionally works with pencil and paper, ruler and t-square, borrowing some design conventions from architecture, and some from engineering. Very often the designer will be working over the architect's own plans, adding a further layer of information to them. But this may not be enough. The completed presentation design needs to work on three levels. Firstly, it needs to show physically where (and what) light fittings are to be located in the design space. Secondly, it needs to show how the electrical circuits supplying and controlling power to the fittings will be placed and organized. Thirdly, the designer needs to present the client with an idea of how the final lighting effects will appear.

For the first two kinds of design there are recognized conventions in architecture for plans, elevations and axonometrics, and

in electronics there are a series of symbols for switches, lamp units, junctions boxes and so on. However, there is no universally acknowledged coding for different lighting units, and so every design drawing will need to be accompanied by a key to identify the precise product used on the plan. Some examples of how this can be done are shown on the illustrations on these pages.

Models and mock-ups

Showing the client what the final scheme will look like has always been more difficult. Whereas an architect can produce a perspective drawing to show the intended appearance of a building, lighting is too ephemeral an effect to be well conveyed by drawing. The conventions of drawing - shading, painting or colouring - interfere with the clear presentation of lighting effects. A second alternative was to build a model, another method regularly used by architects to understand and explain the form of a building. However, architectural models present problems of scale. It is very difficult to find scaled-down light fittings that accurately reproduce the colour and intensity of light from full-size fittings, on the one hand. On the other, architectural models are mainly made from foam block or balsa wood, and these do not share the same surface characteristics as proper materials, so that lighting effects on them will be incorrect.

The drawbacks of models can be compensated for by full-scale testing - for example the effect of a particular light source on a particular surface can be tried with samples, and an idea of the overall effect calculated from that. In the case of a large project, like a multistorey office building, a

Internally lit model for a building in Pennington Street, by architect Rick Mather (above), and the full-size model of Pier 4A (right). Models on display by architect Stanton Williams at the RIBA (facing, above) and a model by Paul Atkinson Associates, using the Horizon lighting gantry.

single office or office suite can be fully installed, to check lighting effects and designs, before the final office is fitted out or occupied, or even, if the project warrants it, before the project is built. This was the case with Pier 4A where, as we have seen, a complete section of the interior was built, full size, not only to examine lighting solutions but also for ergonomic, security and safety tests.

Computer-aided design

The advent of computer-aided design systems was not immediately useful to lighting designers for visualizing lighting effects. However, the ability of the computer to calculate and draw plans, elevations and circuit diagrams did ease the workload in design offices. At the same time architects began to become familiar with the new technology and use it increasingly for drafting, and for visualizations. One of the other main ways in which the lighting designer can now benefit from computer systems is by using them to check light levels in an overall design.

A two dimensional or three dimensional model can be constructed on the computer, based on the plans and elevations of the building, and individual luminaires placed in it. The software then places a net of sensors over the space, which record light levels at positions chosen by the operator (floor, eye or desk level, for example). This is presented as a contour map on screen, or can be printed out. From this the designer can adjust the position, intensity and angle of each luminaire and lamp until the required solution is reached. The programme can then be used to generate a positional plan for the

HOW TO USE LIGHT

placement of the fittings, a list of all the material elements required, and a note of the total wattage, colour appearance and colour rendition of the lamps involved.

As to surface effects, when the reflectance, colour saturation and hue of the wall-coverings is entered, the programme will also calculate the exact colour appearance of the walls when lit. (An accurate printout of the appearance is not possible with current technology, but colour values can be checked on the Munsell scale.) Second-generation CAD systems, which have much higher resolution graphics, are beginning to allow lighting designers to see their handiwork on screen in a reasonable approximation of reality, but higher graphics are normally at the expense of exact measurement. Better graphics mean better presentations to the client, but less exact data for the engineers to work on. We prefer to concentrate on getting the engineering aspects correct, with our present system.

Parametric CAD

Third generation systems hold out the possibility of properly-lit architectural walk-through, in which the designer, architect and client can "visit" the space in real time, either through a computer screen and touchball, or interactively through a virtual reality helmet and glove. Third generation systems will also have computational advantages, as purely graphic elements in CAD programmes are replaced by object-based data units, called parametrics, which will contain far more than just spatial information. Current systems such as Sonata are already object-based, and allow not only visualization of a design but multiple layers

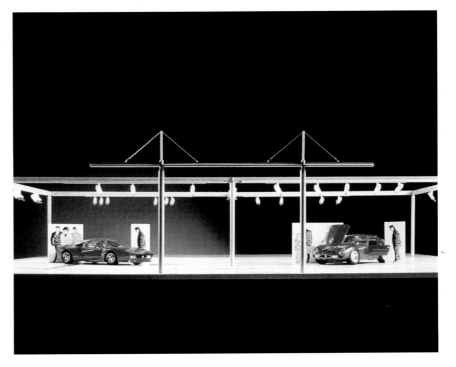

For example, when asked to design a gentleman's club for a client in Tokyo, the architect Julian Bicknell consulted Giuliano Zampi over the exterior lighting. Studies made on Sonata enabled changes in the colour and positioning of exterior floodlights to be made rapidly. The appearance of the building in daylight could be studied, as could the effects of different weather conditions.

of analysis as well, through the database of information within each screen object.

Giuliano Zampi is an architect and computer expert who was one of the team working on the development of Sonata. He is also particularly interested in lighting, and in the problems of presenting lighting designs. Because the Sonata programme uses a parametric rather than a graphic language, it is possible to create a database for each luminaire which will include not only the location and physical size of the lamp, but also its ambient drop-off, polar curves, and colour values. On screen, therefore, the light projected by each luminaire can be seen, and the effect of moving the position of the light, or changing the colour or intensity of it, also viewed. Since these changes mare made via the database, there is no need to redraw the whole image on each occasion, the computer does this automatically.

This kind of technology is becoming increasingly available to lighting designers, and helps take some of the guesswork out of planning a project. But bear in mind that all presentation systems - printouts, screen images, colour transparencies - all distort colour values slightly in the process, and change the scale between original and image, in the same way that a model does.

CAD projects: George Street, Edinburgh

One recent project on which computer studies were used was a new building at 10, George Street, Edinburgh. The architects, Reiach and Hall of Edinburgh, had the task of inserting a modern building into the historic context of Georgian Edinburgh, onto a street replete with classical facades. The intended use of the building as offices required an open and flexible floor plan, while the facade had to have the gravitas of its surroundings. The excellent views across George Street and over northern Edinburgh to the Firth of Forth, and from the back of the building across Edinburgh Castle suggested using glass extensively.

According to Neil Gillespie, the designer of the building, "we wanted to avoid the facile solution of adding classical detail to the facade, and decided instead for using the grid of the design itself to create layers of density and interest." The roof-line is stepped to accommodate the transition in height between adjoining buildings, and the elements of the facade increase in complexity floor by descending floor. The window framings and proportions, and the stepping of the bays, echo the main grid. At ground level, this complexity shows in the bronze screen, glass wall and stone-faced plinths that meet the pilasters framing the main bays, which in turn break up the even run of the upper storey facade. The lighting solution to the facade was therefore implicit in the whole design concept.

It was to uplight the pilasters from the plinths, stressing this with modern bronze tubular torchéres at first floor level, the latter feature echoing classical torcheres elsewhere in the street. Concord were asked to advise

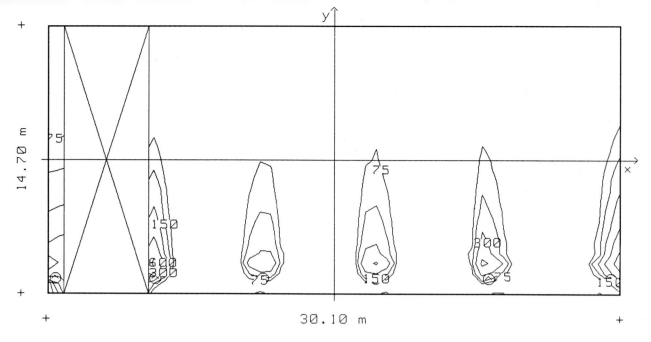

Computer-generated images of light patterns on the exterior at George Street (above), enable the designers to check the lighting effect before installation. The Concord computer programme was designed by Alex Stockmar. It calculates light values in lumens from specific sources at different levels in a room both in plan and in perspective, using coloured lines to map different levels of light. On page 94, the images show light levels in the Concord office at eye level and worktop level (top) and the light distribution in perspective (bottom).

on the choice of lighting, and used computer studies to decide which luminaire and lamp best fitted this task. We decided that a warm white light would best enhance the colour of the stonework, and that 150 watt narrow beam HST lamps in IP55 aluminium reflectors would give the best distribution with most of the light falling in tongues of white up the pilasters, and little overspill. Six luminaires are in place, four along the base of the pilasters, and one on each side of the main entrance porch. The lighting softens the lean modern lines of the building, without taking away from its rigour. The outer faces of the pilasters are marked by neat vertical scorings, a sober enrichment highlighted by the lighting.

The design has been praised for its honesty of architectural approach and the success of the lighting. Also, the building, completed at the end of 1992, is already let, a tribute to its fitting so aptly into the Edinburgh financial and business district.

CAD projects: Concord offices

Another typical use of computer simulations is in office interiors, where it is important to achieve an even level of lighting over working surfaces. Concord's own mezzanine offices posed us a particular challenge. The floor area of 195 square metres is three times as long as it is wide, and the ceiling height was only 2.45 metres, accentuating the long, narrow space. Structural support pillars for the upper floors also break up the space. The room is bounded down one long side by angled exterior windows and on the other, which is slightly shorter, by a high balcony overlooking the showroom, itself with a planar glass exterior wall to the full height of ground floor and mezzanine. The open-plan working area includes desks, drawing boards, and computer workstations.

Our first decision was to create a central corridor down the centre of the room, just to one side of the pillars, and then to place the computer section on the end wall, where problems of glare on the screens could be minimized. The metal ceiling cladding was painted white, to give it a high reflectance, and the pillars were left with their concrete exposed. Grey carpeting and pale grey furniture completed the overall colour scheme. It was also preferable to use uplighting as the main light source, as this would give a feeling of greater space vertically. Tracked or hanging uplighters, however, would negate this effect, and so we chose Toltec uplighters mounted on custom made stands.

Putting the design onto the computer allowed us to calculate the positions and light levels that would create an average illuminance of 450 lux on the work surfaces

The Concord drawing office.

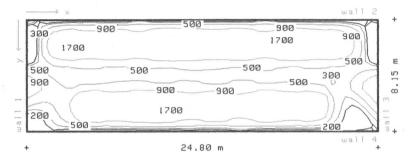

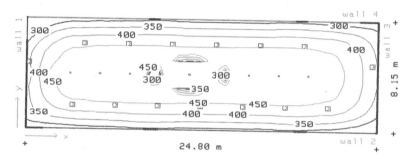

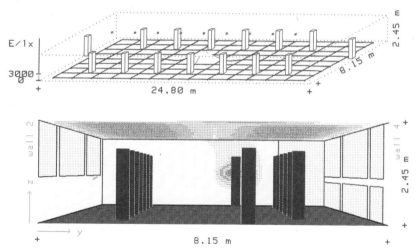

in the main area, falling off to 350 in the computer area. Accurate colour rendition in the main area is also important. The optimum solution was to use fifteen uplighters, thirteen with double pods giving a total 340 watts each, and two singles of 170 watts. These were arranged in two rows, of six and seven each, parallel to the long walls, with the two singles at either end just behind the centre line.

We also wanted to use light to define the corridor running through the work area, and to emphasize the contrast we used coned silver downlighters, mounted in the ceiling. These were fitted with narrow beam lamps. To heighten the contrast further, lamps with different colour appearances were used in the two fittings. Fifty-five watt tungsten halogen lamps, producing a warm white light, were fitted in the downlighters, while high pressure discharge lamps of 170 watts, giving an intermediate, and so slightly cooler white, were used in the uplighters. Small downlighters were also fitted in the obelisks supporting the uplighters, but these were wholly decorative, and hardly contribute to the overall lighting scheme.

This may seem a complex solution to a relatively simple problem, but we had to bear in mind the difficulties presented by the low ceiling and the mix of different activities. So using a custom-made support for the uplighters, for example, was justified by the need to create a distinctive feature in a rather bland room. But if uplighting had not been the right lighting solution, then a customized unit would have been doubly wrong.

HOW TO USE LIGHT

Design into place

These case studies have moved on from the design phase to the final design, so let's build on this by looking at two more projects, ones in which the designers faced similar problems though in rather different settings.

St Nicholas Centre, Sutton

An American architect once described the shopping mall as "the icon of modern living" to which a critic replied that he must have meant "eyesore". Whether we like them or not, malls have become part of the suburban landscape, and their architectural vocabulary is now well-established: it consists of an open central area with shops flanking it, on two or more levels, and often with a glazed roof. At each level pedestrian walkways allow access to the shops on one side, and a general view over the central area on the other. For the architect such spaces provide a challenge of flow - getting customers and their cars in and out of the building's middle, and getting the shopkeeper's services and supplies in around the outside. They also pose problems of identity. If the mall itself is too bland, the customer does not feel welcomed, and goes elsewhere. If the mall is too imposing, has too much presence, the customer may be alienated, and the individual shops will lose their sense of place within the whole.

The lighting designer's contribution to the first problem is concerned with practicalities - assuring that directional signs are properly lit, and that levels of lighting in service areas are safe and secure, for example. In the case of lighting for the customer, the lighting design can make a crucial contribution. A good lighting design will link the different parts of the building together, to

A general view of the St. Nicholas Centre, showing the glazed roofs and two storeys of shops above the main floor. The architect of the centre was Barry Wilde. This daylight view is a reminder that the appearance of lighting fixtures can sometimes be of paramount importance.

A further view of the Centre (above right) showing the metal halide luminaires suspended on arms below the roof, and where the design style matched the special wallmounted fittings (above) which contain both compact fluorescent and dichroic lamps.

help establish its identity, while also allowing the individual shops in the complex to be easily identified. One way to do this is to concentrate the main lighting in the central well, allowing the shop's own lighting to light the walkways.

At the St Nicholas Centre in Sutton, a long, narrow, pedimented roof, fully glazed, covers three levels of shops. The interior walls were to be finished in white, as were the glazing bars. The balustrades and walk-

HOW TO USE LIGHT

way panels were in glass, with dark brown handrails. The resulting architecture, with broad, plain columns rising to the roof level, was cool and almost severe. The lighting designer, Barry Wild, decided to create a special luminaire for the building. This one-off design would help generate a visual identity for the mall, as well as enriching the details of the architecture. And light could be used to soften the harsher features of the building.

The St Nicholas luminaire consisted of a low voltage lamp and a fluorescent. It took the form of two narrow tubes set at a slight angle to the parallel, with a central decorated panel, surmounted by an etched glass fan. The low voltage lamp is fitted behind the central panel. Its light grazes the wall surface, putting the fitting itself into relief. The compact fluorescent is mounted behind the glass fan. The fan diffuses the light, creating a halo effect. Like modern flambeaux, these fittings are placed beside doors and at other key points.

On the different levels lighting had to be supplied between the walkways and in the central well. Here two solutions were found. For the walkways, uplighters were designed which used the same visual features as the special luminaires - paired columns in black tubing, with a tubular crossbar supporting the light fitting. This was a shallow dish into which a compact fluorescent was installed. The dish was in sandblasted glass like the fan on the luminaires. Below these fittings on the walkways, where the edge of the walkway projected into the central space, circular fluorescents were installed, also behind circular sandblasted glass discs. To project light into the central well, metal halide downlighters were suspended from arms

projecting from the side walls. The design feel of the arms echoed the design of the special luminaires, and this was completed by placing circular floating glass dishes below the downlights.

The designers' aim at the St Nicholas Centre was to create coherence. The four main types of light fitting used shared a common design vocabulary. This heightened the sense of continuity between one part of the building and another. And by passing most of the light across or through etched glass, a further uniformity was obtained.

The same design style is carried through into the uplighters on the walkways (above).

The main hall of the Spanish Glass Museum, used for displaying industrial items: ambient illumination was the key here.

Spanish Glass Museum

Glass was also important, but in a very different way, in a recent museum project in Spain. Through our Spanish distributors, we were invited to advise on the lighting of a new museum of glassware being established in a converted building - itself a former glass factory. The first challenge was the building itself. The main rooms were barrel-vaulted, with roughly plastered stonework ceilings and walls. The floors were stone flags. The textures were very interesting, particularly the exposed opus reticulatum on the arches between the rooms. These spacious, cool rooms were to be used as display areas for glass production techniques - kilns and moulds and grinders - from the period of hand-blown glass to contemporary industrial methods. Part of this section was also set aside as an information area. The glassware itself was to be exhibited in a series of smaller rooms, on a different level. These had dark wooden floors and sloping, beamed ceilings with exposed woodwork. Within them display cases would contain the glassware.

The barrel vaulted ceilings, pierced by blind window arches, clearly deserved to be lit in their own right, and this could provide sufficient general illumination at floor level. The windows had been blocked off, and so artificial light was the necessary solution. We therefore placed uplighters on the cornice level, either between the window bays or on the squinches of the cross-vaults. The uplighters were fitted with metal halide lamps, which gave a clear bluish white light, reflected off plaster and stone, over the whole area. This arrangement emphasized - almost dramatized - the architecture, and

HOW TO USE LIGHT

The information display area, adjacent to the main hall (above left) and the special display area for glass objects (above right). The same tracked system was used in all three areas to provide continuity and versatility.

provided continuity between the display and information areas.

To light the exhibits themselves, and the information displays, the Lytespan suspended tracked system was installed. The tracks, each rectangular in plan, were suspended just below the cornice level. Narrow and wide beam spotlights could be placed at any point on the track, though generally they were placed in the sides parallel to the main walls, with paired spots in the centre of the other sides indicating the central aisle through the exhibits. These spots were fitted

with PAR30 lamps, which provided a warmer light. The contrast between the exhibits and their architectural frame was therefore maintained by the slightly different colour of light used in each area.

Lighting display cases is often difficult, because of the problems of reflections off the outside glass surface and refraction of light passing through it. Lighting display cases full of glass objects was doubly difficult. One solution is to mount the lighting system in the top of the display case itself. This creates the impression of the objects being lit from

A display case in the special display area, using internal and external lighting for the cases., and a general view of the main apse, as fully lit.

with floodlights. This, of course, created some reflections in the display cases, so, making a virtue of necessity, we had the wooden floor highly polished. By adding to the number of reflections, we hoped to reduce their influence on the visitor.

Using track in the exhibition area created a further link, both in terms of appearance and colour of light, with the other display areas, even though the adjacent architecture was different. The main aim of this museum design was to provide a lighting system that would show the very different works on view to best advantage, that would enrich the architectural setting, and which would be coherent to the visitor. Changing the light colour between different areas was one way of achieving this. Fitting a tracked system also allowed for modification of the lighting arrangements either for changes in the layout or for new displays to be incorporated.

within, and can be put to good effect. The drawbacks are that there is not much flexibility in the lighting position, and that care needs to be taken in selecting lamps that do not generate too much heat where delicate art works or objects are concerned. Recent developments in fibre optic lights have proved extremely useful in this context, as we shall see later.

In the glass museum the chosen option was to light the cases internally, with dichroic pin spotlights illuminating the objects. Heat was not a problem, and we were able to put the reflectivity and transparency of the glass objects themselves to work. This solution did not, however, provide quite enough light in the room as a whole, and a further suspended track running around the whole room just below the roof line was installed and fitted

HOW TO USE LIGHT

This chapter studies finished lighting designs in place, grouped by their contexts. We look at lighting for exhibitions (an area in which Concord plays an important role), lighting in retail settings, lighting in offices, lighting in restaurants, and finally the special category of lighting for exteriors. Where relevant, the specific lighting options used are mentioned in the caption text, as they have been in earlier captions.

You should bear in mind that these examples are not formulae: they have been chosen to show different approaches, not fixed solutions. While each different context has its own specific needs and requirements, as we shall see, it is crucial not to typecast lighting designs, not to say to oneself "it's a shop (or an office, or a bar) so I've got to do it this way."

I try and get round this in my own work by looking at the space from the human angle. What are people going to do in this space? Are they going to meet each other, are they going to work together, are they here to enjoy themselves, are they going to look at things, or are they going to be alone? Is this a space in which people will stay for long periods of time, or just be passing through? What kind of mood or atmosphere do I need to create for the people using the space? Does it have to be calm and tranquil (in a workspace, for example) or does it need to relax people (in a restaurant) or excite and interest them (as in a shop)? What kinds of interaction are going to go on between the users themselves and the space? Compare, say, a room for meetings and one for presentations, or a dentist's surgery and a hairdressers salon. The good lighting solution is the one which enhances the human activity in the space, as much as the one which makes

SAMPLES & EXAMPLES

Corridors can often be dull, but careful lighting can enliven them (above).

Trade fairs
These two stands are from
Euroshop 93 in Dusseldorf.
The Hindsgaul stand is lit
on the interior with
tungsten halogen low
voltage spotlights and
projectors suspended from
a ceiling frame ((this page:
exterior, facing, top left).

Lighting for exhibitions

The importance of lighting at trade fairs, at exhibitions and in museums and historic buildings cannot be overestimated. The overall aim is in each case the same, to encourage the public to look at what is on display, but between the different applications priorities are different.

Trade fairs are often held in general-purpose arenas which house other events in the year. Sometimes stands are built by the organizers, sometimes exhibitors provide their own. In either case we are talking about temporary structures, however large or sophisticated some of them might be. The task at a trade fair or temporary exhibition is to provide lighting that is attractive and efficient. There will be stands nearby

competing for the public's attention, so a bit of drama or bravura is no bad thing.

Often the rules regarding light levels, electricity supply, fire and safety risks are set out by the fair organizers, and these need to be carefully observed. Also, setting up and dismantling needs to be planned carefully, as often only a limited time is available for each stage.

The Concord Horizon lighting gantry was developed with trade fair work in mind. It provides a framework not only for light fittings but also for display systems, and is designed to be easily mounted and dismounted as an independent item.

In lighting art exhibitions the cardinal rule is that the work of art comes first, the light-

The Chris stand, also at Euroshop, (left and below) used 75 watt low voltage lamps with rotating parabolic reflectors to create changing pools of light focused on the mannequins themselves.

ing designer second! There are also security requirements, and conservation aspects to be considered: works of art with some painted, drawn or dyed surfaces can suffer rapidly and irreversibly if exposed to too much light, both in the visible and ultraviolet spectrum. Too much heat is also undesirable for certain materials (paper, cloth and animal products particularly), which often need to be displayed in controlled conditions of temperature and humidity.

One of the most important developments in lighting for exhibition work, and especially within display cases, is fibre optic illumination. Fibre optic lamps work by passing a beam of light down the myriad tubes of a strand of glass fibre, where it is

A product design exhibition is somewhere between a trade fair and an art exhibition: at *The Inventive Spirit* (right) the bluish light from metal halide spotlights accentuated the colour and styling of the Olympic racing bicycle, and so was acceptable.

Art Exhibitions

These two examples (below and facing, below) show different ways to light paintings . One uses an even flow down the wall, the other adds pools of light for contrast. In both cases the light level on the actual painting is the same, and both are from the Hayward Gallery, London.

diffused at the end. This has various advantages: firstly the light is cool at the emitting end, thereby reducing heat near the objects illuminated. Secondly, it is possible to achieve accurate colour rendition at low light levels, which is important with light - sensitive objects such as photographs or watercolours. Thirdly, the lamp unit and starting end of the fibre-optic cable can be at a distance from the object displayed, or the display case. Thus changing lamps, in the case of failure, can be done away from the object, and, for instance, with a display case, without opening the case and so creating a risk to security and climatic control, as we have seen.

The security aspects considered, the designer also needs to work closely with exhibition staff in deciding how work is to be lit, from what positions and angles it is to be seen. Reflections on glass surfaces and overlighting are two of the principal dangers. Overlighting can flatten the appearance of a paintings surface, or obscure the details of a carving. Lighting can also play a role in guiding the visitor through an exhibition, by creating a visual interest in the exhibition space itself, and by highlighting key works

Sculpture is never easy to light, and Giacometti's work (above) is no exception. The idea was to create an even wash of light from the recessed uplighters behind the false backing wall.

Fragile, valuable and sensitive objects often need display cases. These can also help preserve ideal atmospheric conditions (temperature and humidity especially) but pose problems of reflection for the lighting designer. The three examples compare different approaches, at an exhibition of ancient glass (below), at the Hayward Gallery, with a standing display (right) in Cologne's Museum of Germano-Roman Antiquities, and at the Asprey's stand Grosvenor House Antiques Fair (far right).

Lighting museums and historic buildings poses many of the same problems of security and lighting level control as lighting art exhibitions. However, within the more permanent context of a museum the lighting designer has greater freedom of manoeuvre in planning lighting effects, a freedom that can be exercised further in the case of a historic building, where suggesting to the visitor part of the house's history can also be a legitimate part of the designer's task.

Permanent collections

Standing exhibitions, such as those of the Burrell
Collection in Glasgow (above), or the Bristol Maritime
Museum (top left) or the Musée du Louvre (below left)
impose the same constraints in terms of fidelity to the
work displayed and of conservation, but allow the
designer to work to a more elaborate scale.

The Gas Hall Gallery

Learning, culture and leisure all interface within a modern museum. Informing and entertaining the visitor have equal value with conservation. When the nineteenth century Birmingham Gas Hall was converted into an exhibition hall, lighting was one technique particularly used to heighten the visitor's enjoyment. In the entrance hall (above) light falling from above invites the visitor to move up to the exhibition space, created by architects Stanton Williams under the original glass and cast-iron roof. Here the changes between the central area and surrounding original building are exploited by different light levels.

The lighting in the exhibition area comes from track mounted low voltage control spots suspended from the main ceiling, with additional uplighters between pediment and coving in the side-walls for greater effect (left). The original features blend in well with the new fabric of the building (above.)

The traditional department store (such as Fenwick's at Brentford, above) requires overall, even and, above all, flexible lighting. Here compact fluorescent downlighters have been used in a regular pattern over the ceilings. This allows counters and display cabinets to be rearranged, repositioned and refurbished to suit the store's needs without excessive relighting.

The shop-window has become an increasingly important area, promoting not only the goods on offer but the ideas of the store. This display by Paul Dyson for Harvey Nichols ran concurrently with a Salvador Dali exhibition (above).

Lighting for retail

The retail marketplace is an intensely competitive one, and as such has developed a wide range of techniques and strategies for attracting the customer's business, whether in food retailing, giftware, fashion or, increasingly, in services. The range of functions a shop needs to encompass has also grown: stock control, order checking, payment processing, packing, security, customer movement - the list seems endless. Lighting can play a useful role in bringing these elements together into a working whole, as well as playing a key part in establishing the ambience of a shop interior (along with the architecture, decorative features and goods on display).

Successful lighting design for retail therefore depends on the designer getting a very full brief from the client, as well as bearing in mind the technical considerations of access, safety and visibility. The examples

This fruit and vegetable shop in Switzerland (above, left) also uses cool and even light to present the wares naturally. In this case fluorescent tubes above parabolic low brightness louvres were chosen. With perishable goods the heat emitted by light-sources is one consideration: another is the appearance of the food displayed. In Harrods Food Hall (above right) the fish display is lit with narrow beams from tungsten-halogen spotlights. A similar problem is elegantly solved by Julia Wykeham at Harvey Nichols meat counter.

Three retail spaces where clarity was clearly part of the design brief: the Wellgate shopping centre in Dundee (right) is built under a glass serre, and special uplighters, as shown were designed to maintain light levels in the evening or on short winter days. For the MUJI shop a menswear and general retailer, the designer chose a lighting system that emphasised the clear wood colours used in the interior design (above right) . At a grocery store in Kashiwa City a custom-built track supports low-voltage spotlights, to create a bright and welcoming atmosphere (above).

on these pages have been chosen to empha-sise how great, and how inspiring, the challenge of lighting retail spaces can be.

The increasing sophistication of shop design sees fittings such as marbled floors and wall textures, greenery and careful signage as basics. The lighting, as in this fashion interior (above), needs to be equally smart. A cooler design approach is seen in another shop (right).

The office refectory (above) used clear colours (black and browns) with a simple layout to create a practical but welcoming space. The tracked fluorescents are set out in a rectilinear grid, which emphasises the clear layout, as well as lighting it effectively. The same effect can be seen used in the café (right), where the simple device of lights set into the ceiling coffers is rendered more effective by architectural details such as the round window, and fittings such as the minimal wooden stools by the bar. A more complex approach is found in a German arts centre restaurant (right).

Lighting for restaurants

Everyone has favourite eating places; neon-bright American diners with Cadillac-pink and ice-blue formica, leather chairs and damask tablecloths in dim-lit clubbery, chattering pavement cafés in the Paris sunshine, chandeliers and engraved mirrors in a Victorian pub, or nouvelle cuisine and bleached hessian. And everyone knows that after the food, it's the mood and the atmosphere of a restaurant that counts. Here, needless to say, lighting plays its part.

Designing the lighting for a restaurant is always an exciting job. The lighting designer needs to work closely with the client and the interior architect or designer, notably in looking at the materials to be used for floor and wall coverings, and for tables, chairs and fittings. Particular attention is needed on the times of day the restaurant will be open, and the contribution daylight will make. A frequent problem area is light spillage, from kitchen and entrance doors, another is balancing the light within

114

For the coffee British Museum (right) a special light fitting was designed, which fits as a collar around the supporting pillars. Mini-spot downlighters supplement the general uplighting projecting above and below the collar. It was very important to balance the colour of light from the two different sources.

Architects and interior designers increasingly make use of unusual materials or wall surfaces to create an individual atmosphere for bars or restaurants. In the bar area above, for example, the large steel sheet behind the bar plays a keen role in heightening the contrast with the pale wood of the chairs, tables and the bar itself. In another bar unit (right) the dark sheen of metal and marble, combined with the columnar form of the unit, made downlighters washing over the curved surfaces the obvious solution.

the restaurant, café or bar with light in adjacent areas, particularly where the restaurant is part of a hotel, or situated in a shopping mall or leisure complex. But even the simplest spaces for eating can benefit from careful lighting design.

SAMPLES & EXAMPLES

The servery (above) was designed by students at the Royal College of Art. While not in use (top right) the lower counter rolls back to form a single piece of furniture. The lighting design therefore needed to work in both situations. This was achieved by uplighting the recessed ceiling area above the back of the unit, and supplementing this with ceiling -mounted downlighters which would light the serving area directly when it was in place.

At the Minema, designed by David Betheim, (right), the upper restaurant level and the bar below not only had to work together harmoniously - the same colours and surfaces being repeated between them, but also be seen as a whole from the street (above).

A doctor's or dentist's surgery (above) may only be a simple one-roomed space, but as a work area it needs careful consideration. Here the solution used has been to place wallmounted uplighters to supplement the daylight filtered through the plain gauze curtains of the windows. A similar combination of uplighting and daylight is found in the office space (above, right) fitted under the roof of a London building. A suspended ceiling can also be used (right) to vary the internal volumes of offices.

Lighting the workplace

As we saw when discussing lighting Concord's own offices, good office lighting can help office efficiency and staff morale. The way a company's offices are lit is also part of the image that the company gives to clients and visitors, and so getting the lighting right benefits the company both ways, both internally and externally.

The lighting designer needs a clear understanding of the kind of work being undertaken in each space to be lit, particularly nowadays if computer screens are going to be used. The movement of people through the office space is also a consideration. We are also now realising that giving people working in offices control over their own lighting is a very positive factor, and any new office lighting plan needs to take this on board.

In the office space there are also technical rules and regulations over providing emergency lighting, over ensuring minimum light levels (in some countries this includes providing access to daylight) that have to be respected. In other workspaces, such as hospitals, factories and schools, these regulations may be stricter, and national lighting federations, such as the C.I.B.S.E. in

Britain and the I.E.S. in the USA, as well as the C.I.E. in the European Union, also publish recommended light levels for different kinds of work. We have already referred to these, and the designer needs to be aware of them.

Parallel to workspaces are public access spaces, such as sports and leisure facilities. Although additional regulations may apply to such spaces, the designer has also in some ways a freer hand in planning their lighting. Because these sites very often interface with work space - as in hotels, airports, courts and libraries, I have grouped the illustrations together.

A library or resources area (above, left) needs good overall lighting, here as designed by Andrew Holmes, used to show off the wood panelling and fittings. Even a traditional boardroom (top) can benefit from additional light through a glass brick wall, and from a variety of surfaces, including the crimson end wall (designed by Gordon Murray for Rank Xerox): in a modern boardroom (above) additional features such as display screens may be needed, and the designer has to take these features into account.

Leisure and work spaces interface in many situations: one such is the hairdresser's shop. The one above is in Yokohama, and the designer has used light very effectively to provide a visual stimulus to the client and a good working light for the hairdresser. The entrance to an executive lounge at Heathrow Airport, designed by XMP (left uses light to combine with the architecture to make a powerful statement.

Transport areas need to welcome visitors, give them clear directions and ensure their safety. In fitting out the airport terminal and the rail station serving it in Okada, Japan, the architect and designer chose a vocabulary of plain forms and strong colours, with a uniform approach to lighting, either downlighting free areas or uplighting the edges of internal structures, to create an overall scheme.

Hotels also require good lighting, and need to create an atmosphere of welcome and comfort for the visitor, be it in a modern setting (above) or one more traditionally fitted into the Art Deco interior of the Park Lane Hotel, London, (above, left).

Some workspaces, in contrast, require excellent technical lighting only, as in the control room (above) for British Telecom, where the users need a clear view of the display screens on the facing wall, or in a recording studio (right), at Metropolis in London.

Lighting exteriors

Around August 15th last summer astronomers predicted a fireworks display in the sky of shooting stars. However, to enjoy this spectacle two things were needed: the sky had to be free of cloud, and the spectator well away from a town or city. The reason for the latter requirement is simply that the increasing use of exterior lighting in cities creates a backwash of light that dulls the appearance of the night sky, and makes the stars - even shooting ones - hard to see. Light from the ground is diffused by dust in the

Longleat House at dusk (top) and at night (above, left). Compare this version with the previous lighting (above, right) where glare from the light completely killed the details of the facade.

atmosphere above cities, and so blots out any view of the incoming starlight.

The main source of light at night is of course street, road and motorway lighting. The aim of this kind of lighting is to provide a safe area for pedestrians or motorists, enabling them to find their way without risk. Such lights are often designed to provide a reasonable amount of light at as low a cost as possible, and so for example on main roads and motorways, especially in Britain, low-pressure sodium lamps are often used. These

SAMPLES & EXAMPLES

Moonlight through trees (above left) has to battle with the backwash of light from urban sources. Uplighting a building at night can be effective, but adds to the overspill of light, which is why designers chose downlighting fluorescent track lighting for this railway station (above). Effective lighting at night does not require massive light sources, as can be seen in the illuminated pergola (left).

The Tent at the Tate, designed by the architect Alan Stanton, was a temporary structure put up by Windsor and Newton for a series of events related to drawing and painting. Although the Tent was mainly used by day (above, right) it was a deliberate decision to light it at night, using both metal halide and white son light sources.

give an even yellowish light, in which shapes can be distinguished and warning and information signs read, but under which colour values merge into uniforms greys and blacks. Other British street lighting systems give an orangey, off-white light, and in Europe there is further variety in the colours of light used. The designer needs to be aware of this, and take into account the strength and the colour of any adjacent lighting, and the impact it will have on the building or landscape to be lit.

Street lighting developed from a twin need - the wish for security and a sense of civic pride. Much the same reasoning applies to exterior lighting today. The owner or occupier of a new building wants to ensure

St. Katharine's dock in London (this page) was a former dockside warehouse, used by the ivory trade, and now transformed into offices and flats. The building was lit by Design Lighting Projects, by grazing a wash of tungsten halogen light up the building, and then using pencil spots to pick out the old projecting lifting booms and other architectural features. The whole scheme is reflected in the water of the basin adjoining he dock.

SAMPLES & EXAMPLES

Daylight and dusk shots of the special suite at Heathrow Airport, designed by Michael Manser.

its security (and the security of those working there) and make its presence noticeable. Just as an architecturally important building can reinforce a company's image, so can a building that serves as a visual landmark at night. Today, almost any new building project will include a plan for external illumination.

The illumination of historic buildings is another area which has developed rapidly in recent years. Often this is linked to the tourism and heritage industry, and to a developing awareness of the importance of historic buildings. Once confined to castles and churches, now a whole range of subjects is possible: railway bridges, town halls, almshouses.

A third area might be called nocturnal effect lighting. By this I mean lighting an interior space (a shop window, for example) with its exterior, nocturnal effect in mind. Or lighting the foyer of a hotel to be as welcoming to the evening visitor as the daytime one. This is in a sense a contradiction of the lighting designer's usual approach, which is to banish the night by putting light in its place. Here the designer deliberately exploits the night as a contrasting background to a lighting effect. In designing for an exterior, the same general principles apply as for interiors: balancing light supplied with the space to be lit, respecting the use of the space, being sensitive to the textures and qualities of the space. Not all the same range of lighting solutions are possible: uplighting and wall-washing are the commonest approaches, just because of the restraints on placing light fixtures, which often have to be at ground level and project upwards. The examples on these pages show only a few of the possible

SAMPLES & EXAMPLES

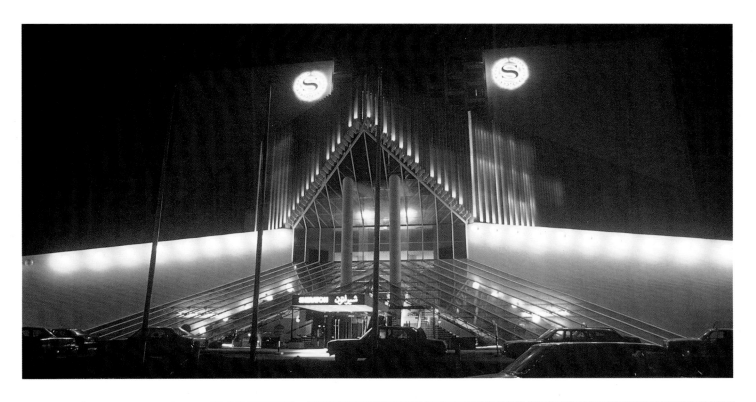

Two views of the Sheraton Hotel in Dubai, the evening exterior shot (above) and an interior shot showing how the interior/exterior lighting relates. Another project with the Sheraton group is the Sheraton Doha (left).

applications of exterior lighting. Fittings for exterior use need additional protection against weather, and increasingly against accidental or deliberate damage, but much the same range of luminaires is available in exterior as in interior forms, together with some special types such as amenity lighting. Particular technologies in exterior lighting include systems sensitive to changing light levels in the surrounding area, allowing for the luminaire to switch on and off according to sunset and sunrise, for example.

Conservation of energy should be among the lighting designer's considerations in every project, but especially in exterior work. In an interior excess light is unfortunate, but it is not wasted, since it is reflected back, in part at least, into the space being lit. Excess exterior light is dissipated into the night sky, wasting the energy that went to produce it. The "light pollution" of Las Vegas is the best known example of this unfortunate effect. And inefficient exterior light does nothing to improve the look of the building being lit, and nothing to improve the general quality of life of users of the building or its surroundings. Excess lightspill from cities at night is a problem for aircraft, a headache for astronomers - and a disappointment for those of us to like to look at the night sky, especially on nights with shooting stars.

Using warm or coloured light at night is not new (the London pub, facing, far left, is a traditional example, though with modern lighting). The lighting for the Hoover building in London (by Imagination) is another example (facing, above): the light colour reflects the original paint scheme. Terry Farrell's London office Building (facing, below) makes a strong silver and pale green statement perched over the River Thames. The Italian example (above) makes effective use of highly sculptural special supports.

IN CONCLUSION

Even lighting designers need to do some home-work! This is the demonstration area at Concord, where architects, interior and lighting designers can see and test a range of fittings in different combinations.

The work of the modern lighting designer seems to become more complex every day. There are always new technologies, new materials, new regulations, new attitudes to understand. In the twelve months in which I have been writing this book, for example, several new luminaires alone have been introduced.

At the same time, not only are the designers' immediate clients, the architects, building managers, and entrepreneurs, becoming more aware of lighting, but the wider public too is also more aware of lighting issues. A lighting design now has to satisfy a much broader clientele than before. And so the range of those who will appreciate a successful design is also broader.

Defining the successful design

So what are the criteria for a successful lighting design? I think there are three broad ones. In the first place a design has to meet the wishes of the clients and users. It has to carry out their brief, respecting their needs and views. It has to respect and follow the relevant building and safety regulations. Because such regulations vary in detail from country to country, and change with time, I have not talked about them in detail here. But understanding and following them must lie behind any decision a designer takes: very often the designer will be the only person on a team to know and understand such requirements fully, and so must be aware of the responsibility involved. Remember, though, that in any case of difficulty, you can always approach the manufacturer of a luminaire or fitting for advice and clarification.

Secondly, a design to be successful must be efficient. This does not merely mean that it be cost-effective, either to install, run or maintain, though these are very important areas in themselves. By an efficient design I mean a design that uses resources as sparingly as possible. 'Green design' in lighting is not always possible: after all the greenest design would be to use daylight only! But the immediate and wider environmental aspects of a design should always be borne in mind. Economy of means should be a guiding factor.

This does not mean underlighting a design. It means achieving a design that creates the desired light effect with just the right resources, neither too much nor too little. It also means designing with the future in mind. Will the lighting system be clear to later users of the building? Can relamping be carried out safely and easily? Can additions or changes be made to the system to allow for changing patterns of use and need? Often, a design that does not look clear and elegant on the drawing board - or even on the computer screen - will normally not be easy to use. So from the start of any design or project the designer should have the ultimate user in mind.

The third criterion is one that is more personal to the designer. It is that every design must, in some way, be innovatory. The design must satisfy the designer's personal values, and take the designer's own work forwards. The satisfaction a designer gets out of a successful design is as important as the pleasure the user gets: but it should not be measured by the number of awards (or even the size of the fee!) The basis of measurement must be the designer's own philosophy of design. The design must, in a word, be an authentic one. It must express the designer's own feelings and values.

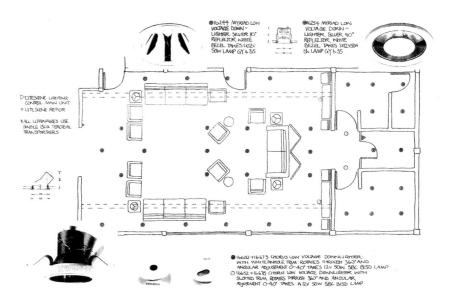

This single room (above) is based on the above plan. Note the conventions and remarks used to define different luminaires (see also page 141).

IN CONCLUSION

The Centre Pompidou in Paris, designed by Renzo Piano and Sir Richard Rogers, with Peter Rice as structural engineer. Concord lit the building when it opened.

[1] From Peter Rice *An Engineer Imagines*, Artemis London, 1994.

The architect-engineer Peter Rice expressed the same sort of idea about his own work. He was a continuous innovator, famous for his work on the Sydney Opera House, on the Pompidou Centre in Paris, on the Menil Gallery in Houston and at the Seville Expo. What he said chimes with some of my own feelings.

'The search for... authentic character... is at the heart of my approach to engineering design. This statement may seem excessive and even frivolous to many engineers. And it is important to emphasize that one should not invent and innovate just for its own sake. Innovation should have a real purpose and be contributing to the project. Nevertheless, some extra design element should be every engineer's objective whatever the project.'[1]

The lighting designer does not always have as wide a field of action as an engineer. But Peter Rice's idea can be transferred to lighting. A good design is one which extends the designer's own vocabulary, and pushes out the envelope of his own design world. A good design will not be the repetition of an existing solution. It may deal with a familiar problem, but the solution will be neater, will incorporate new ideas, or new technologies. It will look forwards, not to the past

The future of lighting

At the beginning of this chapter I mentioned the rate of technical innovation in lighting. If I had to look into the future, I think I would see lighting being challenged to develop in three main areas. These are innovation, specialisation and optimization.

Innovation we have touched on regularly in this book. Two main areas in which lighting technology is changing are in

IN CONCLUSION

the development of new light sources, and in miniaturisation. A good example of the former are the fibre optics. Back in the 1950s, the application of fibre optics in lighting was only a curiosity, in table lamps made from bunches of single waving strands. Now fibre optics, as we have seen, are providing light units that have important and useful applications. Other light sources are also evolving, leading to lamps with even more varied ranges of white colours, with longer lamp life, and with less energy consumption. These developments may not be as visually spectacular as the fibre optic development, but they have important consequences for the lighting designer, in extending the range of choice available.

Miniaturisation is also gathering pace. The first compact fluorescents appeared a few years ago, packing the output of larger lamps into smaller and smaller envelopes. This followed the earlier development of small luminaires with discharge lamps, notably in tungsten halogen. But for compact we can now read miniature. The T2 miniature fluorescent is already available, though fittings to use it are slower coming on stream. Other miniature lamps will surely follow. On the technical side, the twenty-first century is likely to see the increasing use of smaller and smaller units, freeing up the lighting designer's choices in almost every application.

I use the word specialisation in two slightly different senses, for the professional designer and for the amateur. We haven't talked much about the needs of the residential user of lighting in this book, which is intended largely for professionals. But just as awareness of the importance of good lighting is growing in the professional sector, among

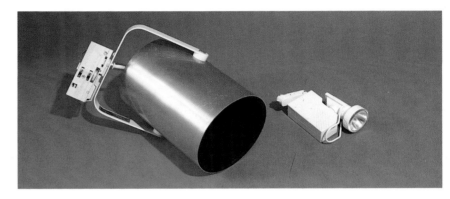

Two examples of new miniaturized products, a downlighter and a spotlight.

Good lighting practice in the home is not difficult: note how the drawing room above is modified by the addition of a ceiling mounted spot, pealights, and opal globes.

architects and interior designers, so the same awareness is spreading to a wider general public. All the principles for good lighting design set out in this book apply, of course, to domestic lighting design. The same effects can be created in the home, the same judgements and rules of design should apply. But what is happening now, and will, I believe, happen increasingly in the future, is that the range of lamps and fittings available "off the shelf" to private buyers will grow both in its choice and in its sophistication.

This can be seen happening in the range of shops and manufacturers selling individual contemporary luminaires to their customers, and this will spread into more and more areas of lighting previously considered only for professionals. If this extension of the market is to continue well and safely - which it should - lighting designers need to be setting even higher standards in their work, which the public is going to be using as a benchmark for their own domestic lighting. Such an expansion of special lighting is a thoroughly good thing. The more people whose lives are well lit, the better!

By specialisation for the professional designer I do not want to suggest or advocate the division of the profession into separate specialised areas (interior and exterior specialists, or those only working on shop design or office design, for example.) That would be an unfortunate and a retrograde step. The discipline of lighting should be set as widely as possible, as the principles that govern good lighting design do not vary from one task to another.

Rather, specialisation I take to mean the increasing opportunity for lighting designers to specify customised or short-run

136

Angled ceiling downlighters used in a traditional bedroom setting (left), and wallwashing downlighters in a small bathroom (below).

IN CONCLUSION

New technologies create new opportunities for lighting designers, but the supreme test is to engage and satisfy the needs and wishes of the user.

series of luminaires for special tasks or situations. This is coming about for two reasons. Firstly, the manufacturers of light fittings are increasingly able to produce special fittings in cost-effective small quantities. At Concord we have seen this part of our business developing considerably in the last few years, and some examples of such 'specials' have appeared in this book. Secondly, the budgets for large-scale architectural projects are now often sufficiently large to meet the design costs of specials, and the growing number of architects who take an active interest in the quality of light in their buildings, and whose knowledge of lighting is becoming more thorough and profound, will surely encourage the wider use of specials in the future.

In citing optimization as a key future development I have in mind the need for designers to think in the widest and best terms about their designs and their impact. We are all aware of the growing energy crisis, and the need for the efficient - minimal, even - use of scarce natural resources. Overlighting - and the consequent overuse of energy - is already a thing of the past in the best lighting practice, and it must be banished completely. Technology, as we have seen, is moving to help the designer meet this challenge. The change in terminology from "low voltage" to "low energy" lamps and systems is not merely cosmetic. It did herald a new approach by industry in looking at the environmental aspects of their work. But much of industry is very often led by costs, and designers need to do their part to push industry forwards, and insist on environmentally-aware criteria for lamps, fittings and lighting designs. This is a challenge that, as we all know, becomes more pressing and

IN CONCLUSION

more urgent every day.

At the beginning of this book I mentioned Goethe's remark that "the golden tree of actual life springs up ever green." Two hundred years after Goethe wrote, his claim for the all-embracing view of science and nature, the necessary "greenness" of life, is still valid, and even clearer and more important to us today than even it was to him. To face the problems for human society posed by our increasingly powerful and dominant technologies, while working within technology itself, is a major task for the creative designer today. After several decades of expansion, innovation and consumption, we are now beginning to understand the true nature of our industrial "success". This, I believe, is the most important challenge facing the lighting designer today, to work to find solutions to the problems of the overuse of energy and resources, while still remaining aware of the positive benefits the wise use of technology can bring.

Lighting touches the quality of our lives in myriad ways. Street lighting can make our towns and cities safer and more welcoming, office and factory lighting makes our places of work healthier and happier, shop and home lighting makes our leisure more relaxed and enjoyable. The true end product of any lighting design, however minor or major, is the pleasure, comfort and enjoyment it gives the final users. This human end to the lighting equation is what makes my work as a lighting designer an endless and rewarding challenge.

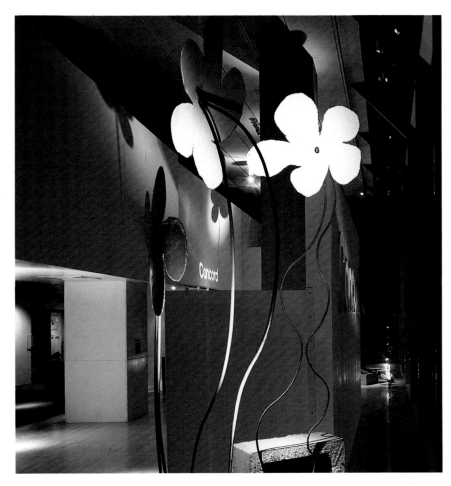

IN CONCLUSION

GALLERY OF LAMPS

For each lamp type the following information is given: classification, former description, wattages, type, initial lumen output, colour rendering group and lamp life in hours.

Class	Description	Wattage	Type	Lumens	CRG	Life
A60	(GLS)	60-100 w,	Incandescent	700-1,350 LM,	1A	1000
A65	(GLS)	150 w,	Incandescent	2,160 LM,	1A	1000
A80	(GLS)	200 w,	Incandescent	2,990 LM,	1A	1000
A-CS60	(CS)	40-60 w,	Incandescent	330-500 LM,	1A	1000
A-CS65	(CS)	100 w,	Incandescent	1,200 LM,	1A	1000
G120	(GlobeGLS)	60 w,	Incandescent	580 LM,	1A	1000
D45	(Round)	40 w,	Incandescent	375 LM,	1A	1000
CO26	(Pigmy)	15 w,	Incandescent	100 LM,	1A	1000
R50	(ISL)	25-40 w,	Incandescent	290 LM,	1A	1000
R80	(ISL)	40-75 w,	Incandescent	440-640 LM,	1A	1000
R95	(ISL)	100-150 w,	Incandescent	920-1,700 LM,	1A	1000
R125	(ISL)	150 w,	Incandescent	1,750 LM,	1A	1000
PAR 38	(PAR 38)	60-120 w,	Incandescent	650-1,740 LM,	1A	2000
PAR 56	(PAR 56)	300 w,	Incandescent	3,000 LM,	1A	1000
QT9	(TH G4)	5-20 w,	Tungsten halogen	60-350 LM,	1A	2000
QT12	(TH M32)	50-75 w,	Tungsten halogen	900-1,500 LM,	1A	2000
QT12	(TH M73)	100 w 24 v,	Tungsten halogen	2,200 LM	1A	2000
QT15	(TH B15d)	150-250 w,	Tungsten halogen	2,5-4,200 LM,	1A	2000
QT30	(TH GY9)	300 w,	Tungsten halogen	5,000 LM,	1A	2000
QT31	(TH E27)	150-250 w,	Tungsten halogen	2,5-4,200 LM,	1A	2000
QT-DE12	(Linear TH)	100-500 w,	Tungsten halogen	1,350-9,500 LM,	1A	2000
QR-CB51	(Tru-Aim)	20-50 w,	Tungsten halogen	350-900 LM,	A	2/3000
QR-CB51	(Dichroic)	20-75 w,	Tungsten halogen	350-1200 LM,	1A	2/3500
QR-CB35	(GR TH)	20 w,	Tungsten halogen	320 LM,	1A	2000
QR38	(SBC TH)	20 w,	Tungsten halogen	350 LM,	1A	2000
QR58	(SBC-GR)	50 w,	Tungsten halogen	900 LM,	1A	2000
QR70	(MR TH)	50-75 w,	Tungsten halogen	900-1,300 LM,	1A	2000
QR111	(PAR 36)	25-50 w,	Tungsten halogen	450-900 LM,	1A	2000
TC	(compact)	5-11 w,	Fluorescent	250-900 LM	1B	5000
TC-D	(compact)	10-26 w,	Fluorescent	600-1,800 LM,	1B	5000
TC-L	(compact)	18-36 w,	Fluorescent	1,2-2,900 LM,	1B	7500
TC-DD	(2D)	16-38 w,	Fluorescent	1,0-2,850 LM,	1B	5000
T26	(fluor.)	18-70 w,	Fluorescent	1,35-6,650 LM,	1B	7500
T-R29	(circular)	22 w,	Fluorescent	1,100 LM	1B	7500
T-R32	(circular)-	40 w.	Fluorescent	2,800 LM,	1B	5000
TC-SB	(SLT)	9-25 w,	Fluorescent	1,000 LM,	1B	5000
T2	(new)	6-13 w,	Miniature fluor.	310-860 LM,	1B	9000

(Not illustrated)

Class	Description	Wattage	Type	Lumens	CRG	Life
HME	*(MBF)*	50-125 w,	H.I.D.*	1,9-6,200 LM,	3	9000
HME	*(MBF)*	250-400 w,	H.I.D.	13,-22,000 LM,	3	9000
HSE	*(SON-E))*	50-70 w,	H.I.D.	3,5-5,600 LM,	4	9000
HSR	*(SON-R)*	70 w,	H.I.D.	4,500 LM,	4	9000
HSE	*(SON-E))*	150-400 w,	H.I.D.	15,-50,000 LM,	4	9000
HST	*(SON-T))*	50-70 w,	H.I.D.	3,1-5,500 LM,	4	9000
HIT	*(HQI))*	35-150 w,	H.I.D.	2,4-12,000 LM,	1B	6000
HIT-DE	*(HQI-TS))*	70-250 w,	H.I.D.	5,5-2,0000 LM,	1B	6000
HIT	*(HQI-MBI))*	250-400 w,	H.I.D.	12,-33,000 LM,	1B	6000
HIE	*(M100))*	75-100 w,	H.I.D.	5,-8,000 LM,	1B	6000

*H.I.D. High intensity discharge

Lamp Shapes

Incandescent lamps

Tungsten halgen lamps

Fluorescent lamps

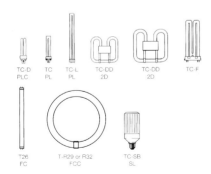

High intensity discharge lamps

FURTHER
READING

There are a number of titles for general readers, at different levels, as listed below.

Gardner, Carl & Hannaford, Barry: *Lighting Design* (Design Council, 1993)

Pritchard, D.: *Lighting* (Longman, 1990)

Sorcar, Profulla: *Architectural Lighting for Commercial Interiors* (John Wiley, 1987)

Sudjic, Deyan: *The Lighting Book* (Mitchell Beazley, paperback, 1993)

Watson, Lee: *Lighting Design Handbook* (Mc Graw Hill, 1990)

Magazines are also an important source for worked examples of lighting schemes and for new technical developments: *Design*, *Architect's Journal*, *Blueprint* and *World Architecture* are the most interesting UK sources, but the interiors sections of *Elle Decoration*, *House and Garden* and *The World of Interiors* are also useful.

The *International Design Yearbook* (Laurence King, 1985 to date) contains an annual review of new lighting products, selected by a major international designer or architect..

Other general books that I have consulted in preparing this one include Tom Porter's *How Architects Visualise* (Studio Vista, 1983), Richard Gregory's *Mind and Science* and Peter Dormer's *Design Since 1945* (Thames & Hudson, 1993).

For technical reference the Lighting Design Federation in the UK produces a guide to *Interior Lighting Design*, as well as a range of other technical publications. Also valuable in the technical domain is the *CIBSE Code for Interior Lighting*.

Manufacturer's catalogues are also a useful guide to available products and systems, and their technical aspects.

AUTHOR'S ACKNOWLEDGEMENTS

Many friends and colleagues, both at Concord and in the lighting business worldwide have contributed to this book, sharing expertise, lending illustrations, listening to ideas. Architects, designers, engineers, lighting companies, and photographers, I am grateful to them all.

Absolute Action,
Lorenzo Apicella,
Paul Atkinson Associates,
Austin Smith Love,
BAA,
Bega,
David Betheim,
Julian Bicknell,
Mark Boyle,
Howard Brandston,
Nigel Coates,
Peter Cook,
Crescent Lighting,
Andrew Dempsey,
Rashhed Din,
Dula Concord (Spain),
Paul Dyson,
Terry Farrell,
Barry Gasson,
Neil Gillespie,
Wilson Goff,
Richard Gregory,
Nicholas Grimshaw,

Julie Hodges,
Andrew Holmes,
Michael Howells,
Indoor, Holland,
Alan Irvine,
Eva Jiricna,
Paul Jodard,
John Johnson,
Simone Kessellheit,
Mary Fox-Linton,
Litecube, Japan,
Sam Lloyd,
David Loe,
Andrew Logan,
Christopher Lucas,
John Lyndon,
Eric Maddock,
Michael Manser,
Paul Marantz,
Rick Mather,
John Miescher,
Mike Morrison,
Gordon Murray,

Jo and John Peck,
Tom Porter,
Julian Powell-Tuck,
Regent (Switzerland),
Lynda Relph-Knight,
Alan Stanton,
Alex Stockmar,
Susan Stockwell,
Harry Swaak,
André Tammes,
Technolyte (Italy),
Bob Thompstone,
Thorn Lighting,
Bob Venning,
Jacques Villain,
Lorna Wain,
Barry Wilde,
Paul Williams,
Terence Wordgate,
Julian Wykeham,
XMP Design,
Giuliano Zampi,
Luciano Zucchi

Finally, very special thanks are due to Vera White, for her efficiency and endless good humour, and to other colleagues at Concord and S.L.I. Germany, Belgium, France and Australia.

INDEX